The 2nd International Forum on Biosecurity

Summary of an International Meeting
Budapest, Hungary
March 30 to April 2, 2008

Katherine Bowman, Jo L. Husbands, and Ben Rusek, *Rapporteurs*

Committee on International Outreach Activities on Biosecurity

Development, Security, and Cooperation

Board on International Scientific Organizations

Policy and Global Affairs

NATIONAL RESEARCH COUNCIL
OF THE NATIONAL ACADEMIES

THE NATIONAL ACADEMIES PRESS
Washington, D.C.
www.nap.edu

THE NATIONAL ACADEMIES PRESS 500 Fifth Street, N.W. Washington, DC 20001

NOTICE: The project that is the subject of this report was approved by the Governing Board of the National Research Council, whose members are drawn from the councils of the National Academy of Sciences, the National Academy of Engineering, and the Institute of Medicine. The members of the committee responsible for the report were chosen for their special competences and with regard for appropriate balance.

This project was supported by the Carnegie Corporation of New York and the Alfred P. Sloan Foundation. Additional funds were provided to support the Forum by the InterAcademy Panel on International Issues, the International Union of Microbiological Societies, and the International Union of Biochemistry and Molecular Biology. Any opinions, findings, conclusions, or recommendations expressed in this publication are those of the author(s) and do not necessarily reflect the views of the organizations or agencies that provided support for the project.

International Standard Book Number-13: 978-0-309-12829-2
International Standard Book Number-10: 0-309-12829-3

Additional copies of this report are available from the National Academies Press, 500 Fifth Street, N.W., Lockbox 285, Washington, D.C. 20055; (800) 624-6242 or (202) 334-3313 (in the Washington metropolitan area); Internet, http://www.nap.edu.

Suggested citation: National Research Council. 2009. The 2nd International Forum on Biosecurity: Summary of an International Meeting, Budapest, Hungary, March 30 to April 2, 2008. Washington, D.C.: The National Academies Press.

Printed in the United States of America

THE NATIONAL ACADEMIES

Advisers to the Nation on Science, Engineering, and Medicine

The **National Academy of Sciences** is a private, nonprofit, self-perpetuating society of distinguished scholars engaged in scientific and engineering research, dedicated to the furtherance of science and technology and to their use for the general welfare. Upon the authority of the charter granted to it by the Congress in 1863, the Academy has a mandate that requires it to advise the federal government on scientific and technical matters. Dr. Ralph J. Cicerone is president of the National Academy of Sciences.

The **National Academy of Engineering** was established in 1964, under the charter of the National Academy of Sciences, as a parallel organization of outstanding engineers. It is autonomous in its administration and in the selection of its members, sharing with the National Academy of Sciences the responsibility for advising the federal government. The National Academy of Engineering also sponsors engineering programs aimed at meeting national needs, encourages education and research, and recognizes the superior achievements of engineers. Dr. Charles M. Vest is president of the National Academy of Engineering.

The **Institute of Medicine** was established in 1970 by the National Academy of Sciences to secure the services of eminent members of appropriate professions in the examination of policy matters pertaining to the health of the public. The Institute acts under the responsibility given to the National Academy of Sciences by its congressional charter to be an adviser to the federal government and, upon its own initiative, to identify issues of medical care, research, and education. Dr. Harvey V. Fineberg is president of the Institute of Medicine.

The **National Research Council** was organized by the National Academy of Sciences in 1916 to associate the broad community of science and technology with the Academy's purposes of furthering knowledge and advising the federal government. Functioning in accordance with general policies determined by the Academy, the Council has become the principal operating agency of both the National Academy of Sciences and the National Academy of Engineering in providing services to the government, the public, and the scientific and engineering communities. The Council is administered jointly by both Academies and the Institute of Medicine. Dr. Ralph J. Cicerone and Dr. Charles M. Vest are chair and vice chair, respectively, of the National Research Council.

www.national-academies.org

COMMITTEE ON INTERNATIONAL OUTREACH ACTIVITIES ON BIOSECURITY

Members

Michael T. Clegg (Chair), University of California, Irvine, Irvine, CA, USA
Gail H. Cassell, Eli Lilly and Company, Indianapolis, IN, USA
Hernan Chaimovich, University of São Paulo, São Paulo, Brazil
Roderick J. Flower, William Harvey Research Insitute, Queen Mary, University of London, London, UK
David R. Franz, Midwest Research Institute, Frederick, MD, USA
Andrzej Górski, Polish Academy of Sciences, Warsaw, Poland
Indira Nath, Blue Peter Research Centre, Hyderabad, India
Barbara A. Schaal, Washington University, St. Louis, MO, USA
Leiv Sydnes, University of Bergen, Bergen, Norway

Staff

Katherine Bowman, Program Officer, Board on International Scientific Organizations
Kathrin Humphrey, Christine Mirzayan Policy Fellow
Jo L. Husbands, Senior Project Director, Program on Development, Security, and Cooperation
Ben Rusek, Associate Program Officer, Committee on International Security and Arms Control

Preface and Acknowledgements

The 2nd International Forum on Biosecurity, held in Budapest, Hungary, on March 30–April 2, 2008, represents the efforts of a number of individuals and organizations, over the last five years, to engage the international community of life scientists in addressing how to reduce the risk that the results of their work could be used for hostile purposes by terrorists and states. The participants who gathered in Budapest were already engaged in this challenging task, and, therefore, the focus of the meeting was on what had been accomplished and what challenges remained. There was no attempt to achieve consensus, since there exist real and important differences among those involved concerning the appropriate policies and actions to be undertaken. But there was a serious effort to identify a range of potential next steps, and also an effort to identify opportunities where international scientific organizations could make substantive contributions and offer their advice and expertise to policy discussions. The Forum's presentations, discussions, and results are summarized in this document.

The Forum also presented an opportunity to continue collaborations and partnerships developed over the years and to forge new ones. The U.S. National Academies provided the services of the conference secretariat, but many individuals contributed to the Forum's planning and implementation. We were fortunate to have five important international scientific organizations as co-conveners of the Forum: the InterAcademy Panel on International Issues (IAP), the InterAcademy Medical Panel (IAMP), the International Union of Microbiological Societies (IUMS), the

International Union of Biochemistry and Molecular Biology (IUBMB), and the International Union of Biological Sciences (IUBS). In addition, the members of the committee appointed by the National Research Council (NRC) under the chairmanship of Michael Clegg, Foreign Secretary of the National Academy of Sciences (see p. v), provided advice about the agenda, speakers, and participants and served important roles during the meeting. The Forum co-conveners, in particular the members of the Biosecurity Working Group of the InterAcademy Panel—Li Huang from the Chinese Academy, Sergio Pastrana from the Cuban Academy, Gabriel Ogunmola from the Nigerian Academy, Koos van der Bruggen from the Royal Netherlands Academy, and Nick Green and Neil Davison from the Royal Society also provided advice and suggestions and took on key tasks during the Forum. Beyond their valuable substantive suggestions, the IAP, the IUMS, and the IUBMB provided funds to support the travel of participants from developing countries. Working with all these colleagues was a privilege.

We also benefitted greatly from the support of the Hungarian Academy of Sciences, which served as the host of the 2nd Forum. The members of the Office for International Cooperation—Janos Pusztai, director, Katalin Hajos, deputy director, and Judit Szász, program manager—were endlessly helpful and exceptionally gracious hosts. Ms. Szász performed wonders to help some of our participants obtain their visas and we would like to express our gratitude and theirs.

We also would like to express our deep appreciation for the contributions of Kathrin Humphrey, who worked on the project as part of her service as a Christine Mirzayan Policy Fellow. Her superb organizing skills, endless patience with the myriad details of an international meeting, and thoughtful contributions to the development of the program had a great deal to do with the success of the meeting. We were fortunate to have her as a colleague.

Members of the NRC committee and the leaders of the co-convening organizations served as chairs of the plenary sessions, an important task that was much appreciated. We also wish to thank the chairs and rapporteurs of the three working groups—Leiv Sydnes and Alastair Hay, David Franz and Neil Davison, and Angelo Azzi and Ralf Trapp. They helped each of the working groups achieve substantial results and their presentations to the final plenary sessions were the foundation for the summaries of the working groups that we have prepared. The statements made in this summary are those of individual speakers or working group members and do not necessarily represent positions of the National Academies, the organizing committee, or all workshop participants.

We would like to extend our special thanks to Ambassador Georgi Avramchev of the Permanent Mission of the Republic of Macedonia to the United Nations Office at Geneva and Chair of the 2008 Meetings of the Biological Weapons Convention (BWC). Ambassador Avramchev attended the entire Forum and gave an informative plenary address; he also participated actively in the working groups and plenary discussions. The attention that the BWC intersessional process has given to dual use issues and the roles and responsibilities of scientists have contributed enormously to the efforts to engage national and international scientific organizations in biosecurity issues.

Finally, we wish to acknowledge the contributions of all the participants in the Forum. Their engagement in the topics and willingness to share experiences and ideas were essential to the success of the meeting. We have attempted to capture at least a portion of their contributions in this summary, but we cannot do justice to the breadth and variety of what they provided.

This report has been reviewed in draft form by individuals chosen for their diverse perspectives and technical expertise, in accordance with procedures approved by the National Academies' Report Review Committee. The purpose of this independent review is to provide candid and critical comments that will assist the institution in making its published report as sound as possible and to ensure that the report meets institutional standards for quality and objectivity. The review comments and draft manuscript remain confidential to protect the integrity of the process.

We wish to thank the following individuals for their review of this report: David Friedman, Institute for National Security Studies and Israel Academy of Sciences and Humanities; Katsuhisa Furukawa, Japan Science and Technology Agency; Robert Mikulak, U.S. Department of State; Kathryn Nixdorff, University of Darmstadt; Alan Pearson, Center for Arms Control and Non-Proliferation; Graham Pearson, University of Bradford; and Carrie Wolinetz, Federation of American Societies for Experimental Biology.

Although the reviewers listed above have provided many constructive comments and suggestions, they were not asked to endorse the contents, nor did they see the final draft of the report before its release. Responsibility for the final content of this report rests entirely with the authors and the institution.

Katherine Bowman, Jo L. Husbands, and Ben Rusek
The National Academies

Contents

Appendixes

1

Background

INTRODUCTION

From March 30 to April 2, 2008, more than 80 people from 31 countries and from 6 international organizations took part in the 2nd International Forum on Biosecurity (Forum) in Budapest, Hungary.[1] The Forum was cosponsored by the InterAcademy Panel on International Issues (IAP), the InterAcademy Medical Panel (IAMP), the International Union of Microbiological Societies (IUMS), the International Union of Biochemistry and Molecular Biology (IUBMB), and the International Union of Biological Sciences (IUBS). The Hungarian Academy of Sciences was the host of the Forum, and the U.S. National Academies (NA) served as the conference secretariat.

The Forum reflected a growing awareness that, while the rapid developments in the life sciences[2] offer great benefits, they also pose the risk that the knowledge, tools, and techniques that enable these advances might also be used to cause deliberate harm. The Forum brought together organizations and individuals active in the field of biosecurity to discuss

---

[1] Appendix B contains a copy of the agenda and a list of participants. Almost all of the individual presentations made in the plenary sessions and working groups are posted on the U.S. National Academies Web site at: *http://www7.nationalacademies.org/biosecurity/2nd%20International%20Forum%20on%20Biosecurity.html.* Accessed on December 10, 2008.

[2] "Life sciences" is a broad category that includes agricultural sciences, biological sciences, and the health sciences. In addition, there is some overlap with the physical sciences (e.g., biochemistry in chemistry and biophysics in physics) and engineering (e.g., bioengineering or biomedical engineering).

the roles and responsibilities of the international scientific community in fostering policies to address these risks, in order to promote both continuing scientific progress and greater international security. More specifically, the meeting addressed the challenges and opportunities to:

- Build a culture of responsibility within the science community regarding biosecurity, through education and awareness raising, codes of conduct, and other mechanisms;
- Identify standards and practices for research oversight from the review of proposals through the conduct of research, publication and communication, and the range of approaches to achieving their widespread adoption;
- Provide scientific advice to governments and international organizations and develop the role of the science community in global governance.

The participants came from all over the world because the life sciences are a genuinely global enterprise, and thus any policies must include international as well as national measures.

As described later in this chapter, the Forum in Budapest was the second international meeting organized by international scientific bodies to address these issues. The first International Forum was held in Como, Italy, in March 2005.[3] The Forum is thus part of a broader process of engagement by the scientific and policy communities in considering biosecurity issues.

The structure of the Forum was intended to encourage discussion and to identify common ground where possible. Working groups were organized to run through the course of the Forum, so that ideas could percolate and develop. These groups, organized to reflect each of the Forum's goals, became the heart of the meeting. On the final morning, plenary sessions offered the opportunity to report back and to discuss the results of the working groups.

The 2005 Forum in Como did not produce a final report, but this time the organizers wanted a written record. The sponsoring organizations agreed that the 2nd Forum would not produce recommendations, and that the final report would be only a summary of what occurred during the meeting. However, each of the working groups held during the Forum was encouraged to make suggestions for next steps and needed actions. These were reported to and discussed in the final plenary, and

[3] The agenda, list of participants, and copies of the presentations from this Forum can be found at: *http://www7.nationalacademies.org/biso/Biosecurity_Forum.html*. Accessed on December 10, 2008.

are included in the final report. Since the Forum secretariat was in the National Academies, the National Research Council (NRC), its operating arm, appointed a committee to oversee the preparations for the meeting (see Appendix A). The planning committee did not participate in the drafting of this summary, which was written by the NRC staff who supported the secretariat, serving as workshop rapporteurs.[4]

The rest of this chapter attempts to synthesize the history of recent developments that provided the context for the Forum. This material was presented by participants throughout the plenary sessions and working groups. Some of the details reappear in the summaries of the presentations and discussions at the Forum, but they are assembled here in one place in hopes of providing a more coherent narrative of events. Chapter 2 then provides a summary of the plenary sessions and discussions, followed by the reports of the three working groups. The final chapter offers a brief summary of the major themes and suggestions for possible actions and next steps that emerged from the discussions.

DEVELOPMENT OF THE ISSUE

Continuing advances in the life sciences over the last 50 years, supported by enabling technologies such as vastly increased computing power, have brought great benefits for health, the economy, and the environment, and promise far more in the future. Along with the hopes, however, have come concerns that the knowledge, tools, and techniques gained through these developments might also be used in state or terrorist pursuit of biological weapons (BW). A frequently quoted warning about the potential risks came in 2000 from Matthew Meselson, a leading figure in the life sciences on issues related to biological weapons:

> Every major technology—metallurgy, explosives, internal combustion, aviation, electronics, nuclear energy—has been intensively exploited, not only for peaceful purposes but also for hostile ones. Must this also happen with biotechnology, certain to be a dominant technology of the coming century? During the century just begun, as our ability to modify fundamental life processes continues its rapid advance, we will be able not only to devise additional ways to destroy life but will also be able to manipulate it—including the processes of cognition, development, reproduction, and inheritance. A world in which these capabilities are widely employed for hostile purposes would be a world in which the very na-

[4] The NRC is part of the National Academies, which also include the National Academy of Sciences, the National Academy of Engineering, and the Institute of Medicine. Created in 1916, the NRC has become the principal operating agency of both the National Academy of Sciences and the National Academy of Engineering in providing services to the government, the public, and the scientific and engineering communities.

> ture of conflict has radically changed. Therein could lie unprecedented opportunities for violence, coercion, repression, or subjugation.[5]

Yet even work in the life sciences that might have the greatest apparent potential for misuse may offer significant benefits as well. The possibility that advances in the life sciences intended for legitimate and beneficent purposes might also be used for malevolent ends is often called the "dual use" dilemma.[6] This is somewhat different from the classic definition in defense and security of dual use that focuses largely on equipment or technology—high-performance computers, advanced materials, "stealth" technology—that could be applied for either civilian or military purposes.[7] This definition reflects increasing attention to developments in science and technology that, although arising largely from academia and the commercial sector rather than from military-related research, raise significant concerns for security. Nanotechnology, microcomputing, and civilian nuclear power are three other areas that are often cited as posing similar dual use issues.

Current concerns about the dual use potential of advances in the life sciences date largely from the beginning of this century and reflect different perceptions—and sometimes sharp disagreement—about the relative risk between the development of national biological weapons programs and the potential for bioterrorism, and between these and other threats to international security. President Yeltsin's admission in early 1992, following years of accusations, that the Soviet Union had maintained a huge clandestine biological weapons program, in violation of the Biological and Toxin Weapons Convention (BWC), came as the revelations of Iraq's efforts to create biological weapons were unfolding in the wake of the first Gulf War.[8] The first World Trade Center bombing in 1993, the Oklahoma City bombing in 1995, and the 1995 Aum Shinrikyo attack in Tokyo with chemical agents, spurred increasing concern with "catastrophic" terrorism.[9] The terrorist attacks of September 11, 2001, and the subsequent

[5]Meselson, M. 2000. The problem of biological weapons. Symposium on Biological Weapons and Bioterrorism, National Academy of Sciences, Washington, DC, May 2.

[6]NRC (National Research Council). 2004a. *Biotechnology Research in an Age of Terrorism*. Washington, DC: The National Academies Press.

[7] Knowledge and skills are included in traditional definitions of dual use, but the emphasis tends to be more on actual items. For a discussion of current debates over dual use, see Reppy, J. 2007. The end of dual use? Implications for export control policy. Paper prepared for presentation at the 48th Annual Meeting of the International Studies Association, Chicago, IL. March. Available at: *http://www.allacademic.com/meta/p_mla_apa_research_citation/1/7/8/8/3/p178830_index.html*. Accessed December 10, 2008.

[8] *Rossiskiye Vesti*. 1992. Interview with President Boris Yeltsin. Washington, DC: Foreign Broadcast Information Service, FBIS-SOV-92-103, May 27.

[9] Carter, A., J. Deutch, and P. Zelikov. 1998. Catastrophic terrorism: Tackling the new danger. *Foreign Affairs* 77(6):80-94.

anthrax letters in October of that year turned those already existing concerns into the highest national security priority, particularly in the United States. In addition, the U.S. response to the perceived threats of bioterrorism included a massive increase in funding for activities of the type most likely to raise concerns, and led some to question whether "defensive" work was becoming increasingly problematic in terms of compliance with the BWC.[10]

In addition to increased concerns about terrorism and state BW programs, a number of articles in scientific journals sparked controversy about whether some research that might be misused should not be conducted, or if conducted, should not be published. Critics charged such publications could provide a "blueprint" or "roadmap" for terrorists or countries seeking to carry out bioterrorism or to acquire biological weapons.[11] Gerald Epstein of the Center for Strategic and International Studies labeled such studies "contentious"; his article was an early review of the issues and policy options then under discussion.[12]

Before proceeding further, it is important to acknowledge that the potential risks of the misuse of advances in the life sciences are not universally accepted. Part of engaging the scientific community in these issues is therefore discussing and debating the nature and seriousness of the risks. On a technical level, some argue that "Mother Nature is the best terrorist" and, therefore, that there exists little reason for terrorists or for less technologically advanced countries to do more than take advantage of the highly dangerous pathogens already abundantly available in nature.[13] On the level of general policy, some consider concerns about bioterrorism to be part of a general U.S. tendency to exaggerate the threat of terror-

---

[10] Miller, J., S. Engelberg, and W. Broad. 2001. *Germs: Biological Weapons and America's Secret War.* New York: Simon and Schuster.

[11] A review of some of the best known articles from that period may be found in *Biotechnology Research in an Age of Terrorism* (National Research Council 2004a, pp. 25-29). An example of the concern in the defense policy community is Zilinskas, R. and J.B. Tucker. 2002. Limiting the contribution of the open scientific literature to the biological weapons threat. *Online Journal of Homeland Security* (December). Available at: h*ttp://www.homelandsecurity.org/journal/Articles/tucker.html.* See also Vogel, K.M. 2008. Framing biosecurity: An alternative to the biotech revolution model? *Science and Public Policy* 35(1):45-54.

[12] Epstein defines "contentious research" as "fundamental biological or biomedical investigations that produce organisms or knowledge that could have immediate weapons implications, and that, therefore, raise questions concerning whether and how that research should be conducted and disseminated." Epstein, G.L. 2001. Controlling biological warfare threats: Resolving potential tensions among the research community, industry, and the national security community. *Critical Reviews in Microbiology* 27:321-354.

[13] For a review of these discussions and debates, see Frerichs, R.L., R.M. Salerno, K.M.Vogel, N.B. Barnett, J. Gaudioso, L.T. Hickok, D. Estes, and D.F. Jung. 2004. Historical Precedence and Technical Requirements of Biological Weapons Use: A Threat Assessment. SAND2004-1854. Albuquerque, NM: Sandia National Laboratories.

ism involving weapons of mass destruction.[14] Other research suggests that absorbing and using new technology may require substantial tacit knowledge that is not easily transferred or acquired by states or terrorists, particularly through published research results.[15]

THE ROLE OF THE SCIENTIFIC COMMUNITY

Responding to the dual use potential of the life sciences is a challenge in which the scientific community has an essential role. The heart of the challenge is developing the mix of policies at the national, regional, and international levels that can mitigate the risks of misuse, while enabling continuing scientific advances and the availability of those advances to all. For many, as illustrated in Figure 1-1, measures to address the risks of BW or bioterrorism are thus best seen in the context of the spectrum of risk to global health and the environment—ranging from chronic disease threats to natural disease outbreaks to the accidental or inadvertent spread of disease to the deliberate use of disease to cause harm.[16]

Sustained effort by the scientific community, drawing on traditions of self-governance and social responsibility, is considered to be an essential component of a broader strategy to respond to the risks of bioterrorism or BW proliferation. In the United States, for example, a number of reports from the NRC have made the aforesaid argument.[17] The scientific community also has an important role to play as advisor to policy makers about trends in science with dual use implications, in assessments of the balance of potential risks and benefits in new and continuing activities, and about the implications of proposed policies for both science and security.

To be effective, responses to the dual use dilemma cannot be confined to national measures. Capacity in the life sciences is diffusing around the world, and thus a meaningful response must include global approaches

---

[14] A detailed and skeptical assessment of this phenomenon related to biological issues may be found in Leitenberg, M. 2005. Assessing the Biological Weapons and Bioterrorism Threat. Carlisle Barracks, PA: Strategic Studies Institute, U.S. Army War College.

[15] Vogel, K.M. 2006. Bioweapons proliferation: Where science studies and public policy collide. *Social Studies of Science* 36(5):659–690; and Vogel, K.M. 2008. Framing biosecurity: An alternative to the biotech revolution model? *Science and Public Policy* 35(1):45-54.

[16] WHO (World Health Organization). 2005. Life Science Research: Opportunities and Risks for Public Health. Geneva: WHO. *WHO/CDS/CSR/LYO/2005.20*. Available at: *http://www.who.int/csr/resources/publications/deliberate/WHO_CDS_CSR_LYO_2005_20/ en/index.html*. Accessed December 10, 2008.

[17] NRC (National Research Council). 2004a. *Biotechnology Research in an Age of Terrorism*. Washington, DC: The National Academies Press; National Research Council. 2004b. *Seeking Security: Pathogens, Open Access, and Genome Databases*. Washington, DC: The National Academies Press; and National Research Council 2006. *Globalization, Biotechnology, and the Future of the Life Sciences*. Washington, DC: The National Academies Press.

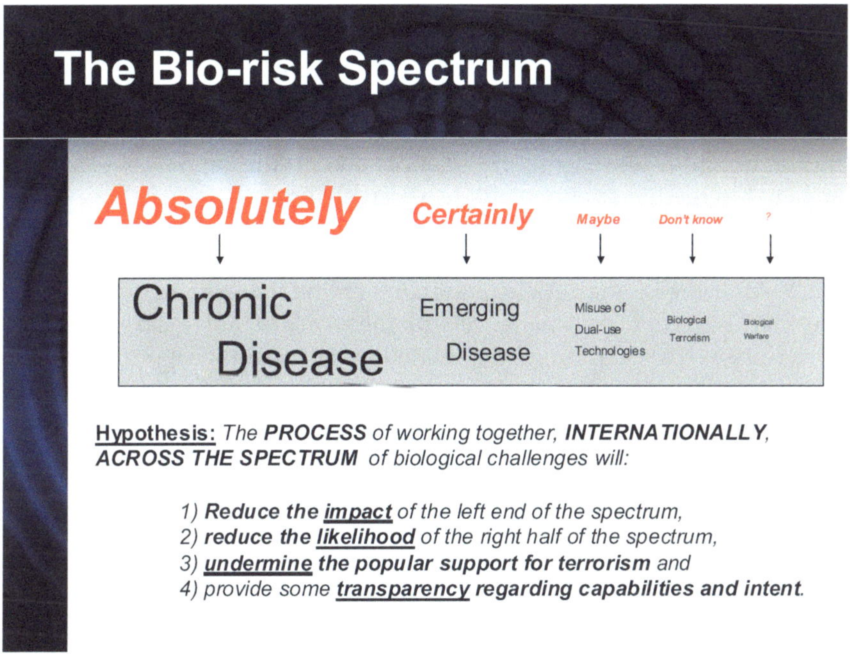

FIGURE 1-1 The bio-risk spectrum.
SOURCE: Franz, D.R. 2007. Challenges and Opportunities. Princeton University, December 18.

as well as national.[18] The failure to undertake compatible international efforts risks, among other things, disrupting the international collaboration that is so much a part of the modern scientific enterprise; scientists sometimes point to the example of the barriers raised by legislation in the United States after September 11 as an example of what should be avoided.[19] Lack of care in the design and implementation of measures to address dual use concerns risks denying access to knowledge and technology in the name of security, or risks driving work into areas where there is less oversight.

Fortunately, an extensive network of national, regional, and international scientific bodies—national professional associations and interna-

[18] Ibid.

[19] See, for example, the results of a survey reported in Fischer, J.E. 2006. Stewardship or Censorship: Balancing Biosecurity, the Public's Health, and the Benefits of Scientific Openness. Washington, DC: Stimson Center. Available at: *http://www.stimson.org/globalhealth/pdf/Stewardship.pdf*. Accessed December 10, 2008.

tional scientific societies, academies of science and medicine, and nongovernmental scientific organizations—offer the opportunity to engage the scientific community. A number of them are active participants in policy debates related to issues of science and society. These bodies are also the most likely and appropriate vehicles to ensure continued commitment to the issues, both within the life sciences community and between those engaged in the life sciences and decision makers.

It must be noted, however, that until recently the life sciences community has not been much engaged in the dual use implications of its work. After the Biological Weapons Convention was signed in 1972, most life scientists had little experience with the issues of biological weapons or bioterrorism; national programs related to biological weapons permitted under the BWC are confined to "prophylactic, protective, or other peaceful" measures. Thus without conscious personal effort or systematic education, very few life scientists working today would have reason to know the details of past offensive weapons programs or have knowledge of the BWC and their responsibilities under that treaty. They also have few connections to the national security branches of government. Moreover, the image of themselves as being engaged in work that is meant only for the benefit of humankind is deeply engrained in the way life scientists view themselves and their role in society. An essential first step is thus raising awareness about the issues within the scientific community.

THE "LANGUAGE BARRIER": ISSUES OF TERMINOLOGY[20]

One of the immediate difficulties that arise in a discussion of the possible potential misuse of the life sciences is the lack of common terms to describe the problem. The term most commonly used, "biosecurity," presents many difficulties. At its most basic, the term does not exist in some languages, or is identical to "biosafety"; French, German, Russian, and Chinese are all examples of this immediate practical problem.

Even more serious, the term is already used to refer to several other major international issues. For example, to many "biosecurity" refers to the obligations undertaken by states adhering to the Convention on Biodiversity and particularly the Cartagena Protocol on Biosafety, which is intended to protect biological diversity from the potential risks posed

[20] "Dual use" is a term that frequently evokes confusion and controversy, but did not receive the same attention in discussions during the Forum. For a review of the multiple meanings of the term, see Atlas, R., and M. Dando. 2006. The dual use dilemma for the life sciences: Perspectives, conundrums, and global solutions. *Biosecurity and Bioterrorism: Biodefense Strategy, Practice, and Science* 4(3):276-286.

by living modified organisms resulting from modern biotechnology.[21] "Biosecurity" has also been applied to efforts to increase the security of dangerous pathogens, either in the laboratory or in dedicated collections; both the World Health Organization (WHO) and the Organisation for Economic Co-operation and Development (OECD) have recently produced guidelines related to practices within this meaning of the term.[22] The term may also have specific national meanings; in New Zealand, for example, the term applies to protecting the island nation from invasive species.

Whatever the problems and limitations with the term "biosecurity," so far no one has been able to develop a better term to describe the policies and practices intended to reduce the risk of misuse of the results of biotechnology. This is the context within which the term was generally used in the international forum described in this report. Adding descriptive adjectives or phrases, such as WHO's use of "laboratory biosecurity," may provide additional clarity.

Biosecurity is also linked to "biosafety." Many of the practices intended to improve laboratory safety and to protect workers and the environment from the accidental or inadvertent release of dangerous organisms have an important relationship to efforts to reduce the risk of deliberate misuse. As will be discussed later in this report, good biosafety practices are part of the foundation for creating a "culture of responsibility" among scientists toward dual use issues. This may be especially true in developing countries where improved biosafety comes as part of building capacity in the life sciences. The distinction between biosafety and biosecurity is primarily that the latter term, as used here, includes the additional consideration of measures to prevent deliberate misuse; biosecurity represents broader societal and ethical issues that are not always included in discussions of laboratory practices to ensure biosafety.

DEVELOPMENT OF SCIENTIFIC ENGAGEMENT

Early Initiatives: Setting the Stage

Many individuals and organizations have played a role in the increasing interest of the scientific community in the dual use dilemma. What

[21] Further information on the Convention may be found at: *http://www.cbd.int/convention/* and on the Protocol at: *http://www.cbd.int/biosafety/*. Accessed December 11, 2008.

[22] WHO (World Health Organization). 2004. *Laboratory Biosafety Manual*, 3rd ed. Geneva: WHO.WHO/CDS/CSR/LYO/2004.11. Available at: *http://www.who.int/csr/resources/publications/biosafety/WHO_CDS_CSR_LYO_2004_11/en/* and OECD (Organisation for Economic Co-operation and Development). 2007. *OECD Best Practice Guidelines on Biosecurity for BRCs (Biological Resource Centers)*. Paris: OECD. Available at: *http://www.oecd.org/dataoecd/6/27/38778261.pdf*.

follows is a rough and necessarily incomplete chronology of some of the efforts, mingling actions by both international and nongovernmental organizations (NGOs).

The fundamental commitment not to use disease as a weapon is embodied in the Biological and Toxin Weapons Convention which was signed in 1972 and entered into force in 1975.[23] As Ambassador Masood Khan of Pakistan, president of the treaty's sixth review conference, stated:

> The BWC has had marked success in defining a clear and unambiguous global norm, completely prohibiting the acquisition and use of biological and toxin weapons under any circumstances. The preamble to the Convention so forcefully states: the use of disease as a weapon would be "repugnant to the conscience of mankind." It captures the solemn undertaking of the states parties "never in any circumstances to develop, produce, stockpile or otherwise acquire or retain" such weapons. With 155 states parties, the treaty is not universal, but no country dares argue that biological weapons can ever have a legitimate role in national defense. Such is the force of the treaty."[24]

In 2002, following the collapse of efforts to negotiate a protocol to the BWC to provide for verification of treaty compliance, the states parties agreed to a series of intersessional meetings before the next full treaty review conference in 2006. Each year focused on a different topic and included both a two week meeting of experts and a one week meeting of the states parties. The topic chosen for 2005 was "content, promulgation, and adoption of codes of conduct for scientists."[25]

The role of codes of conduct for scientists has been a continuing focus of interest with regard to dual use issues. (There are, in fact, several kinds of codes, each with a different purpose;[26] as used here and elsewhere, "codes of conduct" is the commonly used general term.) In addition to

---

[23] UN Security Council Resolution 1540, passed in 2004, adds a further binding international commitment against support for non-state actors seeking to acquire weapons of mass destruction or means of their delivery. Available at: *http://www.un.org/Docs/sc/unsc_resolutions04.html*. Accessed December 11, 2008.

[24] Khan, M. 2006. Preparations and expectations. Presentation to the United Nations General Assembly First Committee. Sixth Review Conference of the Biological and Toxin Weapons Convention: New York: United Nations, October 11. Available at: *http://www.unog.ch/80256EDD006B8954/(httpAssets)/298DFC7CC2CD636BC125720D0045B3C8/$file/First_Committee_BWC_thematic_presentation_slides.pdf*. Accessed December 11, 2008.

[25] Additional information about the topics and contents of other intersessional meetings can be found at: *http://www.opbw.org/* under "Strengthening the Convention."

[26] Rappert, B. 2004. Towards a Life Science Code: Countering the Threats from Biological Weapons. *Bradford Briefing Paper* No. 13. Available at: *http://www.brad.ac.uk/acad/sbtwc*. Accessed December 11, 2008.

the BWC intersessional meeting, as a result of the recommendations of the UN Policy Working Group on the United Nations and Terrorism, the UN General Assembly and Security Council passed resolutions in September 2002 calling on the UN Secretariat to reinforce ethical norms and to prepare relevant codes of conduct for scientists involved in technologies that could produce weapons of mass destruction. The Under-Secretary-General for Disarmament Affairs initially asked the International Centre for Genetic Engineering and Biotechnology (ICGEB) to assist the Secretariat in this task in relation to the life sciences.[27]

In 2002, the International Committee of the Red Cross (ICRC) launched an initiative on "Biotechnology, Weapons, and Humanity," calling for a "web of prevention" to address the risk that technologies from the life sciences could be used for hostile purposes. In addition to a number of proposals for national and international legal measures to support the implementation of the BWC, the initiative recommended including education about risks, rules, and responsibilities as part of the overall ethical training for life scientists. [28]

In January 2003, in response to the controversy over scientific publications mentioned above, a group of editors and authors from some of the leading scientific journals met in Washington, DC, along with experts in security policy and biological weapons. The group drafted a "Statement on Scientific Publication and Security," at the heart of which was the acceptance of responsibility for screening manuscripts to reduce the risk of misuse of scientific information. The statement was simultaneously published in *Science*, *Nature*, the *Proceedings of the National Academy of Sciences* (*PNAS*), and in the journals of the American Society for Microbiology (ASM).[29] The overarching principle accepted by the Journal Editors and Authors Group stated that "there is information that, although we

---

[27] Ripandelli, D. 2005. Building blocks for a code of conduct for scientists, in relation to the safe and ethical use of biological sciences. Presentation to the 2005 Meeting of Experts of the Biological Weapons Convention. Geneva. June 13. Available at: *http://www.opbw.org/*. Accessed on December 11, 2008.

[28] More information may be found at: *http://www.icrc.ch/Web/eng/siteeng0.nsf/html/bwh!Open*. Also see presentation by Coupland on page 23 in this report. Earlier, Graham Pearson coined the phrase "web of deterrence," but he did not address dual use research issues (Pearson, G.S. 1993. Prospects for chemical and biological arms control: The web of deterrence. *The Washington Quarterly* 16(Spring):145-162.)

[29] Journal Editors and Authors Group. 2003a. Uncensored exchange of scientific results. *Proceedings of the National Academy of Sciences* 100(4):1464; Journal Editors and Authors Group. 2003b. Statement on the consideration of biodefense and biosecurity. *Nature* 421:771; and Journal Editors and Authors Group. 2003c. Statement on scientific publication and security. *Science* 299(5610):1149; Fox, J.L. 2003. Bioterrorism threat could make some research too "sensitive" to disclose. *ASM News* 69(3):112-114. Available at: *http://www.asm.org/microbe/index.asp?bid=13147*.

cannot now capture it with lists or definitions, presents enough risk of use by terrorists that it should not be published." The Group indicated that if "the potential harm of publications outweighs the potential societal benefits," manuscripts may be rejected. The statement also notes that publications are not the only place where science is communicated, and that all scientists are responsible for monitoring their communication to maximize the benefits and minimize the risks of their research.[30]

Several journals subsequently adopted formal policies to consider "dual use" and the potential for misuse of the information in the manuscript during the review. Today, the Nature Publishing Group, *PNAS*, the ASM journals, and *Science* have review policies in place, and although the policies are not uniform, they signify continuing concern regarding science and security.

In October 2003, the U.S. National Research Council released a prepublication version of a report that focused specifically on the potential risks of dual use research, *Biotechnology Research in an Age of Terrorism*, often called the "Fink report," after the study's chair, Gerald Fink of MIT.[31] Planning for the project had begun prior to the events of September 11, and prior to the anthrax mailings; but those events gave the report much greater visibility. The report made a series of recommendations, largely focused on enhancing self-governance by the scientific community, but also with a role for federal guidelines and an advisory body modeled on the Recombinant DNA Advisory Committee of the National Institutes of Health.

Although the Fink report focused on the United States, it argued that effective efforts to reduce the risk that biotechnology could be misused would depend on international action.

> Any serious attempt to reduce the risks associated with biotechnology must ultimately be international in scope, because the technologies that could be misused are available and being developed throughout the globe. A number of countries and regional and international organizations are already moving forward to develop programs and policies on aspects of the problem; the initiatives include consultations among the parties to the BWC on best practices for the security and oversight of pathogens and toxins. These approaches must be harmonized and widely adopted in order for them to be effective. Just as the scientific community in the United States must become deeply and directly engaged, the commitment of the international scientific community to these issues is needed to implement the recommendations contained in this report. [32]

[30] Ibid. *Science*.

[31] NRC (National Research Council). 2004a. *Biotechnology Research in an Age of Terrorism*. Washington, DC: The National Academies Press.

[32] Ibid., p. 12.

A number of other important efforts were launched during the same period by nongovernmental organizations (NGOs) or scholars, most often from the United States or the United Kingdom. These include but are not limited to the Controlling Dangerous Pathogens Project at the University of Maryland, educational seminars conducted by Malcolm Dando and Brian Rappert through the University of Bradford, the International Council for the Life Sciences, and the Center for Biosecurity at the University of Pittsburgh Medical Center.[33] Each had a slightly different focus and a more or less explicit policy agenda, but all were concerned in large measure with the issues surrounding what the OECD called "responsible stewardship of the biosciences."[34]

2005 as a Turning Point

One of the challenges for those interested in engaging the international scientific community is the wide array and variety of organizations. There are hundreds, if not thousands, of international scientific meetings every year in all parts of the globe, and a multitude of national and regional groups and groupings. But there are remarkably few genuinely independent international scientific organizations devoted to bringing science to bear on policy issues. This is important because such organizations have a particular advantage in being able to work directly with international and intergovernmental organizations.[35] Many national scientific organizations have a significant international membership (for example, an estimated 30 percent of the membership of the ASM is international), but there are still significant limits on what such national organizations can do in the

[33] See Appendix C for a description of these and other efforts.

[34] OECD (Organisation for Economic Co-operation and Development). 2004. Promoting Responsible Stewardship in the Biosciences: Avoiding Potential Abuse of Research and Resources. Chairman's Summary. Paris: OECD. Available at: *http://www.oecd.org/dataoecd/30/56/33855561.pdf*. Accessed December 11, 2008.

[35] An example is the collaboration that has developed between the Organisation for the Prevention of Chemical Weapons (OPCW), charged with implementation of the Chemical Weapons Convention (CWC), and the International Union of Pure and Applied Chemistry (IUPAC). For example, in 2002 IUPAC held a workshop at the request of OPCW on trends in chemical sciences and technology as input to the first CWC review conference in 2003. The report of that workshop, which was used extensively by the OPCW secretariat in preparing for the review conference, can be found in a special issue of the union's journal (Parshall, G.W., G.S. Pearson, T.D. Inch and E.D. Becker. 2002. Impact of Scientific Developments on the Chemical Weapons Convention (IUPAC Technical Report). *Pure and Applied Chemistry* 74(12):2323-2352. Available at: *http://www.iupac.org/publications/pac/2002/7412/index.html*. The technical papers presented at the workshop are also contained in *Pure and Applied Chemistry* 74(12). A second IUPAC-OPCW workshop on trends is described later in the chapter.

international arena.[36] There are also a number of important international science policy organizations, such as the ICRC and the Pugwash Conferences on Science and World Affairs, but these have a policy agenda and less of a base in the general scientific community.

Beyond the limited number of genuinely international science bodies, none of the obvious candidates among existing organizations—the International Council for Science (ICSU), the IAP, or the IAMP, as described in Box 1-1—had been engaged in issues of science and security beyond the questions of the openness of scientific research and the human rights of science, engineering, and health professionals.

As mentioned above, the topic for the 2005 BWC intersessional meetings was "content, promulgation, and adoption of codes of conduct for scientists." The choice of codes provided an excellent opportunity to encourage scientific organizations to pay attention to biosecurity issues.

The IAP emerged as the primary actor among the three international scientific organizations, but its partnerships with other scientific groups were essential to the broader task of engaging the scientific community. In February 2004, the IAP Executive Committee adopted a Biosecurity Initiative, and formed a small working group under the leadership of the Accademia Nazionale dei Lincei of Italy.[37] Other members of the Biosecurity Working Group (BWG) included the academies of China, Cuba, Nigeria, and the United States. The UK Royal Society became part of the working group in September 2004; later that year the Royal Netherlands Academy of Arts and Sciences took over as chair of the BWG.

The BWG had rather quickly decided to focus its efforts on drafting a statement of principles that could provide the basis for efforts by academies and other science bodies to develop codes of their own rather than attempting to develop a full-blown IAP code of conduct. This reflected in part a view that codes are most effective when those adhering to them have some sense of "ownership," and that this is best achieved when codes come from local or national sources with which people have closer, more direct ties.

---

[36] The special advantages of international status can, of course, be overstated; for example, the Monterey Institute for International Studies has forged a close working relationship with the International Atomic Energy Agency and international and regional organizations work with and support national groups.

[37] The IAP General Assembly had received a proposal in December 2003 from the International Centre for Genetic Engineering and Biotechnology (ICGEB) to collaborate on preparing a code of conduct. It became clear by the fall of 2004, however, that the process needed to create and then gain the endorsement of an IAP statement could not proceed quickly enough to meet the ICGEB's desire to fulfill the UN's request to have a completed code in time for the BWC experts meeting in June 2005. The two efforts, therefore, went forward separately.

BOX 1-1
Some Key International Scientific Organizations

The International Council for Science (ICSU), founded in 1931, is a non-governmental organization representing a global membership that includes both *national scientific* bodies (111 members) and *international scientific unions* (29 members).[a] As its Web site notes: "Because of its broad and diverse membership, the Council is increasingly called upon to speak on behalf of the global scientific community and to act as an advisor in matters ranging from ethics to the environment." Approximately a dozen of ICSU's unions can be considered to be part of the "life sciences"—reflecting the breadth and fragmentation of the field, unlike the single unions for physics and chemistry. ICSU also has a standing Committee on Freedom and Responsibility in the Conduct of Science.

The InterAcademy Panel on International Issues (IAP), founded in 1993, is another global network, comprised of approximately 100 of the world's science academies.[b] It is designed "to help its members develop the tools that they need, in order to participate effectively in science policy discussions and decision making." The current co-chairs are from Canada and China. As one of its major activities, the IAP issues statements that are endorsed by its member academies; the first two statements, on population (1994) and urban development (1996) were timed to coincide with special sessions of the United Nations on those topics.

The InterAcademy Medical Panel (IAMP), launched in 2000, is a global network of 64 academies of science and medicine, committed to improving health worldwide. IAMP activities focus on "institutional collaboration to strengthen the role of all academies to alleviate the health burdens of the world's poorest people; build scientific capacity for health; and provide independent scientific advice on promoting health science and health care policy to national governments and global organizations."

[a]The ICSU Web site is: *http://www.icsu.org/index.php.*
[b]The IAP Web site is: *http://www.interacademies.net.*
[c]The IAMP Web site is: *http://www.iamp-online.org.*

In November 2004, the IAP Executive Committee agreed to a proposal from the U.S. National Academy of Sciences to serve as a co-convener for an International Forum on Biosecurity. The IAMP and ICSU also agreed to serve as co-conveners at approximately the same time.

The International Forum was held in late March 2005, at a conference center in Como, Italy, with the stated goals of:

- Broadening the debate and advancing the awareness in the life sciences and biomedical research communities—and in the international scientific community more generally—about the challenges posed by the dual use dilemma;

- Serving as a major convening and coordinating mechanism to share information about activities already under way or being planned to address biosecurity issues;
- Providing an opportunity for a discussion of these activities, for identifying potential gaps and needs and for how they might be filled, and, in this context, exploring opportunities for future international cooperation and collaboration.

Just over 50 participants from 20 developed and developing countries and from several international organizations took part in the Forum, which included both plenary sessions and day-long parallel sessions devoted to specific topics—codes of conduct, "sensitive" information and publication policy, and research oversight—that enabled in-depth discussion. The IAP draft statement was discussed extensively during the small group session on codes of conduct, for example, and was revised in response to the comments and suggestions. Although the participants were largely scientists identified through IAP or ICSU, participants also included people from a number of the other policy projects on biosecurity, as well as staff from the ICRC, the WHO, and the OECD. [38]

The rules of the Forum precluded reaching formal conclusions or making recommendations—a condition from the IAP and ICSU boards when they agreed to serve as cosponsors—but the ideas generated in the working sessions were summarized and circulated informally among the convening organizations as a basis for their future activities. For example, at its meeting in April 2005, the ICSU Executive Board endorsed further work on biosecurity by the organization and its member unions, thus setting the stage for future engagement and collaboration.

The 3rd Meeting of Experts of the Biological Weapons Convention took place in Geneva, Switzerland in June 2005. As already mentioned, the meeting's focus on codes of conduct had provided an opportunity to encourage scientific organizations to pay attention to biosecurity issues. Moreover, in an important departure from tradition, the chairperson of the meeting offered a variety of professional organizations, NGOs, and outside experts the chance to make brief presentations to the meeting as "guests of the chair," in addition to the usual NGO statements that were part of many such meetings. The chairperson also encouraged member states to include additional experts as part of their delegations. The foreign secretary of the Cuban Academy of Sciences presented the draft IAP

[38] The agenda and participants list, as well as other information and copies of the presentations, can be found at *http://www7.nationalacademies.org/biso/Biosecurity_Forum.html*.

statement.[39] Three of the ICSU unions, as well as ICSU's Deputy Executive Director, also made presentations. Following her presentation and her experience with the meeting, the President of the International Union of Biochemistry and Molecular Biology (IUBMB) convened a working group, which created a code of ethics for the IUBMB; among the obligations to the public, members "will not engage knowingly in research that is intended for the production of agents of biological warfare or bioterrorism, nor promote such agents."[40] The International Union of Microbiological Societies also created a brief code and has urged national affiliates to adopt it, and to craft their own, more extensive codes relevant to local conditions.[41]

The final IAP statement was released on December 1st, just in time for the 2005 States Parties meeting. A copy of the statement, which was formally endorsed by 69 of the then 93 IAP member academies, can be found in Appendix D. The chair of the BWC meeting mentioned the statement in his opening remarks and officially circulated the statement to all the delegations.

In addition to the BWC process, two other important international organizations had also become engaged in biosecurity and dual use issues by 2005. The involvement of WHO and OECD added the elements of global health and economic development to the more traditional security concerns represented by the BWC, and also served to emphasize the need for a mix of policies to ensure that efforts to reduce the risk of misuse also allowed for continued scientific progress. Of particular relevance, the OECD Global Futures Program created a website (*www.biosecuritycodes.org*) to provide information about national and international activities, and the WHO released a background paper, Life Science Research: Opportunities and Risks for Public Health, as an initial step toward increasing engagement in the issue.[42]

Finally, there were important developments at the national level. In

---

[39] Further information on the meeting, copies of many of the presentations, and a copy of the chair's final report, which cites the IAP statement extensively, along with the key points made by the Royal Society and other science organizations, can be found at *http://www.opbw.org/*.

[40] The code can be found on the IUBMB Web site at: *http://www.iubmb.org/index.php?id=155*. The description of its origins may be found at: *http://www.iubmb.org/index.php?id=41#c496*. Accessed December 11, 2008.

[41] *The IUMS Code of Ethics against Misuse of Scientific Knowledge, Research, and Resources* is available at: *http://www.iums.org/about/Codeethics.html*. Accessed December 11, 2008. The code was formally adopted by the IUMS General Assembly on August 10, 2008.

[42] WHO (World Health Organization). 2005. Life Science Research: Opportunities and Risks for Public Health. Geneva: WHO. WHO/CDS/CSR/LYO/2005.20. Available at: *www.who.int/csr/resources/publications/deliberate/WHO_CDS_CSR_LYO_2005_20/ en/index.html*. Accessed December 11, 2008.

particular, the three largest funders of biomedical research in the United Kingdom announced in September 2005 that applicants for funding would now be asked to indicate whether their proposed research had dual use potential, and that dual use considerations would be included in reviews. The joint policy statement from the Biotechnology and Biological Sciences Research Council, the Medical Research Council, and the Wellcome Trust identified a series of agreed actions that the three organizations would implement to raise awareness and to help ensure that any risks of misuse associated with research proposals were considered at the grant application stage.[43]

Developments between 2005 and 2008

The years between 2005 and 2008 saw the international community continue to cooperate on biosecurity issues, although raising awareness among and educating the broad life sciences community remain formidable challenges. This section briefly describes some of the efforts by both independent scientific organizations and international organizations. Other activities and projects are described in Chapter 2 and are listed in Appendix C.

The WHO continued to engage on biosecurity issues by creating a working group and holding a small international workshop in October 2006 on "Life Science Research and Global Health Security." The workshop report recommends the creation of a standing scientific advisory group to counsel the WHO Director-General on biosecurity, including both improved biosafety and responsible oversight of research.[44] WHO has also undertaken a number of collaborative activities, including regional workshops that address both biosafety and biosecurity issues.

In April 2006 the UN Secretary General issued a report calling for a global strategy to counter terrorism. The report covered many aspects of the problem and included the statement: "The most important under-addressed threat relating to terrorism, and one which acutely requires

---

[43] The joint statement is available at: *http://www.bbsrc.ac.uk/organisation/policies/position/public_interest/misuse_of_research_joint.pdf*. Accessed December 11, 2008.

[44] WHO (World Health Organization). 2007. Scientific Working Group on Life Science Research and Global Health Security: Report of the First Meeting. WHO/CDS/EPR/2007.4. Geneva: WHO. Available at: *http://www.who.int/csr/resources/publications/deliberate/WHO_CDS_EPR_2007_4*. Accessed December 11, 2008.

new thinking on the part of the international community, is that of terrorists using a biological weapon."[45] The report then recommended that:

> What we need now is a forum that will bring together the various stakeholders—governments, industry, science, public health, security, the public writ large—into a common program, built from the bottom up, to ensure that biotechnology's advances are used for the public good and that the benefits are shared equitably around the world. Such an effort must ensure that nothing is done to impede the potential positive benefits from this technology. The United Nations is well placed to coordinate and facilitate such a forum, and to bring to the table a wide range of relevant actors. I urge Member States to consider this proposal in the near future.[46]

In September 2006, the United Nations General Assembly passed a resolution creating a UN *Global Counter-Terrorism Strategy*, including a proposal to bring together "the major biotechnology stakeholders, including industry, the scientific community, civil society and governments, into a common program aimed at ensuring that biotechnology advances are not used for terrorist or other criminal purposes, but for the public good."[47] It was hoped that this could become a regular event. Although the transition to a new Secretary General slowed progress, the Secretary General's office is currently developing plans for a major new initiative.

The 6th Review Conference for the BWC held in late 2006 offered an opportunity for some of the international scientific organizations to provide input to the review of the implications of trends in the life sciences for the implementation and operation of the treaty. The Royal Society, in collaboration with the IAP and ICSU, organized a workshop in London to assess the implications of rapid developments in the life sciences.[48] Among its results, the workshop highlighted the importance of monitoring technological developments, such as improved aerosol delivery tech-

---

[45]Annan, K. 2006. Uniting Against Terrorism: Recommendations for a Global Counter-Terrorism Strategy. Report of the Secretary-General. A/60/825. New York: United Nations, p.11. Available at: *http://www.un.org/terrorism/unitingagainst terrorism/contents.htm.* Accessed December 11, 2008.

[46] Ibid., p. 11-12.

[47] United Nations. 2006. The United Nations Global Counter-Terrorism Strategy. UNGA Resolution A/RES/60/288. New York: United Nations, Annex II-11. Available at: *http://www.un.org/terrorism/strategy-counter-terrorism.shtml#resolution.* Accessed December 11, 2008.

[48] Royal Society. 2006. Report of the RS-IAP-ICSU International Workshop on Science and Technology Developments Relevant to the Biological and Toxin Weapons Convention. London: The Royal Society. Available at: *http://royalsociety.org/displaypagedoc.asp?id=22789.* Accessed December 11, 2008.

niques, in addition to purely scientific developments.[49] It also highlighted the increasingly blurred lines among fields such as chemistry and biology in many areas of particular interest and concern, with the emerging field of synthetic biology as a prime example.[50]

One of the decisions made at the 6th BWC review conference in December 2006 was to continue the intersessional meetings until the next review conference in 2011. Reflecting the increasing level of engagement and international interest, the topics chosen for 2008 were:

- National, regional and international measures to improve biosafety and biosecurity, including laboratory safety and security of pathogens and toxins.
- Oversight, education, awareness raising, and adoption and/or development of codes of conduct with the aim of preventing misuse in the context of advances in bio-science and bio-technology research with the potential of use for purposes prohibited by the Convention.[51]

The choice of topics for the 2008 meetings provided another opportunity to encourage further engagement by national and international scientific organizations in convening a meeting directly relevant to their interests.

In April 2007, IUPAC organized its second workshop on trends in chemical sciences and technology for the Organization for the Prohibi-

---

[49] A similar argument is made in the 2006 report from the National Research Council, *Globalization, Biosecurity, and the Future of the Life Sciences* (Washington, DC: The National Academies Press).

[50] This growing field combines elements of biological science, chemistry and engineering into a highly interdisciplinary area of the life sciences. Synthetic biology offers the potential to construct bioengineered microorganisms that might, for example, enable the mass-production of drugs to treat disease, detect and break down toxic chemicals to reverse polluted sediments and water, and generate new energy forms to help solve the energy crisis. There are also substantial concerns, however, regarding the potential for the creation of "dual use" products, either intentionally or unintentionally, that could function as biological weapons or lethal pathogens in the hands of terrorists. For further information and discussion of policy options, see Bügl, H., J.P. Danner, R.J. Molinari, J. Mulligan, D.A. Roth, R. Wagner, B. Budowle, R.M. Scripp, J.A L. Smith, S.J. Steele, G. Church, and D. Endy. 2006. A Practical Perspective on DNA Synthesis and Biological Security. International Consortium for Polynucleotide Synthesis. December 4. Available at: *http://pgen.us/ICPS.htm*. Accessed December 11, 2008. See also Garfinkel, M.S., D. Endy, G.L. Epstein, and R.M. Friedman, eds. 2007. Working Papers for Synthetic Genomics: Risks and Benefits for Science and Society. Available at: *http://hdl.handle.net/1721.1/39658*. Accessed December 11, 2008.

[51] Biological Weapons Convention Meetings Secretariat. 2006. Sixth Review Conference of the States Parties to the Biological Weapons Convention. Final Document. Geneva: United Nations Department for Disarmament Affairs. Available at: *http://www.opbw.org*. Accessed December 13, 2008.

tion of Chemical Weapons. The workshop, held in Zagreb, Croatia, was intended to inform the preparations for the 2nd review conference of the Chemical Weapons Convention in April 2008. A number of topics and themes overlapped with those in the Royal Society-IAP-ICSU workshop on trends in life sciences.[52]

SUMMARY

These are only examples of some of the events that have taken place in the last few years, focusing primarily on international interactions. Other international events, and additional national and regional activities, are described later in the report. Taken together they underscore the increasing opportunities for scientists and scientific organizations to engage with policy makers to develop ways to address biosecurity issues and to provide expert advice about trends in the life sciences, so that policies are based on realistic assumptions.

It is important to recognize that these growing opportunities also pose challenges. Biosecurity is at a relatively early stage of development as an international issue. Because of the complexity of the problem and the importance of reaching diverse constituencies, it is necessary and desirable to have many stakeholders addressing biosecurity through different venues and approaches. In some cases there are genuine disagreements about both the nature and the scope of the problem. Such diversity creates the potential for overlap and duplication of effort, or even for unintentionally working at cross purposes. Multiple approaches are important, but the chances for success are increased if these various efforts communicate and, where reasonable, coordinate their work.

With this as background, we now turn to our account of the 2nd International Forum.

[52] Balali-Mood, M., P.S. Steyn, L.K. Sydnes, and R. Trapp. 2008. Impact of scientific developments on the Chemical Weapons Convention (IUPAC Technical Report). *Pure and Applied Chemistry* 80(1):175-200. Available at: *http://www.iupac.org/publications/pac/80/1/0175/*. Accessed December 11, 2008.

2

Plenary and Working Group Presentations and Discussions

SUMMARY OF PLENARY PRESENTATIONS

Plenary 1: Introduction to the Forum

The plenary discussions at the 2nd International Forum on Biosecurity began with an overview of the issues to be dealt with during the meeting. Roderick Flower (William Harvey Research Institute, Queen Mary, University of London) introduced the Forum and highlighted the goals, objectives, and structure of the meeting. He placed the 2nd Forum into the context of a selected time line of international biosecurity initiatives undertaken since 2001, including release of several influential studies, convening the 1st International Forum on Biosecurity in 2005 and a Royal Society-hosted meeting in 2006, production of the Statement on Biosecurity by the InterAcademy Panel on International Issues (IAP) and the development of further initiatives such as a code of conduct for biosecurity produced by the Royal Netherlands Academy of Arts and Sciences (KNAW) in 2007. The talk highlighted the progress made by the international scientific community in considering dual use issues in the life sciences, the challenges that remained to be addressed, and some of the opportunities that might be presented by the current intersessional process of the Biological and Toxin Weapons Convention (BWC).

Robin Coupland (International Committee of the Red Cross [ICRC]), Ottorino Cosivi (World Health Organization [WHO]), and Alexandre Bartsev (Organisation for Economic Co-operation and Development [OECD]) next formed an introductory panel to provide further context in which to locate the Forum discussions and possible frameworks within which to consider dual use life sciences issues.

Drs. Coupland and Cosivi focused on public health approaches to the potential risks posed by the misuse of products of life sciences and biotechnology, particularly infectious microorganisms. Both the ICRC and the WHO have focused their efforts on analyzing risk factors, effects, and preventive measures. The ICRC has developed the concept of a "web of prevention," in which complementary and interacting efforts from multiple stakeholders combine to offer protection from an outbreak of disease. The presentation drew an analogy to the multiple layers of protection that help prevent or reduce injuries from fires, including smoke alarms, flame-retardant materials, sprinkler systems, and dedicated fire departments. The talk also highlighted the role of the scientific community in fostering a safety and security culture and in raising awareness among scientists of potential risks related to the development, production, and delivery of microbial agents.

Ottorino Cosivi provided Forum participants with a complementary framework used by the WHO in considering global health security. This consisted of a series of interlocking puzzle pieces representing contributions from the areas of ethics, policy, collaborations and support, and laboratory safety and security, which together combined to form the norms, standards, and supporting activities to help manage health security risks. Risk management in this public health context could also be viewed as a matrix in which diverse actors on individual to international levels (including scientific associations, public health laboratories, publishers, funding partners, security communities, and the public) each undertake a range of activities to address components of this puzzle.

As an intergovernmental body, WHO has focused many of its efforts on assisting member countries by working to develop risk assessment methodologies and to produce a tool kit of resources with multiple risk management options. WHO has formed a scientific working group on life science research and global health security that recommended five areas for action: education and training, disease outbreak preparedness, risk assessment methodology development, stakeholder engagement, and capacity building.[1] WHO held a regional workshop in Thailand in December 2007 that recommended further actions by both WHO and its member countries in many of these areas. [2]

[1] WHO (World Health Organization). 2007. Scientific Working Group on Life Science Research and Global Health Security: Report of the First Meeting. Geneva: WHO. WHO/CDS/EPR/2007.4 Available at: *http://www.who.int/csr/resources/publications/deliberate/WHO_CDS_EPR_2007_4.*

[2] Research Policy and Management of Risks in Life Science Research for Global Health Security, Bangkok, Thailand, December 10-12, 2007.

Finally, Alexandre Bartsev spoke to the Forum about how OECD has incorporated biosecurity into several of its recent initiatives. The OECD considers having effective biosecurity procedures to be an enabling tool for economic development and innovation in science and technology. It supports the concept of Biological Resource Centers (BRCs) serving as important repositories of materials and information and has developed best practice guidelines for biosecurity at such Centers.[3] The OECD guidelines address maintenance, access, and distribution of biological materials held in BRC collections; the guidelines include recommendations for undertaking risk assessments and for developing risk management procedures for pathogens with dual use potential. While the OECD currently consists of 30 industrialized nations, additional countries are in the process of accession and this organization has increased engagement with other rapidly developing countries, including China, India, and Brazil. OECD member countries will report to the Council in 2010 on the implementation of the BRC biosecurity guidelines; the OECD considers prospective member countries' implementation of relevant OECD acts and guidelines, including those on BRC biosecurity, as part of the accession process. To assist member countries, the OECD will convene an intergovernmental forum to consider some of the issues remaining with regard to biosecurity risk assessments for microorganisms, including assessment methodologies, how to share and communicate assessments, how to consider local differences in risks and how to balance governance, so as to best enable continued science and technology innovation. Looking to the future, the OECD plans to hold a workshop in 2009 with the U.S. National Science Foundation. The workshop will focus on the biosecurity implication of emerging technologies such as synthetic biology, and will explore ways to incorporate biosecurity practices into the internationally mobile scientific workforce.

Plenary 2: Emerging Life Science and Technology: Challenges and Opportunities for Biosecurity

The second plenary session of the Forum also looked to the future of the life sciences and addressed some selected highlights of recent scientific work. The three panelists for this session were Jason Chin (Cambridge University), Jörg Stelling (ETH-Zurich), and Jane Calvert (Edinburgh University).

[3] OECD (Organisation for Economic Co-operation and Development). 2007. *OECD Best Practice Guidelines on Biosecurity for BRCs (Biological Resource Centers)*. Paris: OECD. Available at: *http://www.oecd.org/dataoecd/6/27/38778261.pdf*. Accessed December 11, 2008.

Jason Chin spoke to the Forum about some of his work in synthetic biology, particularly on designing biological entities with new functions. He explained that the complexity of biological systems has led synthetic biologists to seek ways to reduce some of this complexity and introduce design principles by creating discrete modules to perform particular functions. The hope is that such modules could then be built up into larger assemblies to perform larger and more complicated functions. Analogies can be drawn to building circuits from combinations of resistors and capacitors, and then assembling such circuits into more and more complicated systems until a computer is constructed. Discrete biological modules have already been created to function as on/off toggle switches, oscillators, and edge detectors, for example.

Although synthetic biology is still a fairly new field, it builds on advances in areas such as molecular biology and genetics and in technologies such as rapid DNA synthesis. Improvements in rapid DNA synthesis and assembly and in the fidelity of synthesized and amplified DNA are both important developments for synthetic biology, allowing functional DNA products to be generated from databases or novel sequences. Techniques for generating mutations within DNA and for selecting mutations that lead to desirable phenotypes are also useful. However, the success rate is still very low and there are still limits on the DNA that can be successfully encapsulated into particular cell membrane shells. With further technological developments, the creation of a synthetic bacterium may be only several years away. However, Dr. Chin highlighted the conceptual difference between modifying something that already exists and creating something totally new.

Jörg Stelling continued the discussion by considering the ways in which bioinformatics and computational tools contribute to designing new systems in biology, and the limits of these tools. The desired characteristics of a designed synthetic circuit include robustness (insensitivity to perturbations and noise), stability within the context of a biological system, tunability to control desired properties, and construction feasibility. Dr. Stelling highlighted two large challenges that remain in working with biological systems—the complexity of such systems and the still incomplete characterization of all of the system components and their properties.

Dr. Stelling presented a time-delay-switch circuit as an example. He compared a representation of a simple electronic circuit diagram with the biological version that consists of multiple interacting modules with overlapping functions. Principles of computational modeling and design can produce mathematical equations to describe how to characterize and fine-tune the biological "circuit," but they are complicated by the presence of unknown parameters, lack of quantitative characterization for many components, and nonlinear behavior. Although such model-based rational design of complete biological circuits is feasible in principle, it is currently only possible for simple designs. Rational, computational model-based design in biology poses some challenges that are new compared to traditional engineering disciplines. The expression of biology in terms of mathematical equations scalable to more complex systems remains the key challenge. The presentation ended with a quotation from the statistician G.E. Box that "all models are wrong; the practical question is how wrong do they have to be to not be useful."

Finally, Jane Calvert addressed the Forum to place some of the developments in systems and synthetic biology into the broader context of changes in the life sciences. Dr. Calvert highlighted how systems biology, which studies the ways in which molecules work together in complex systems, opened the path to synthetic biology which aims to create and build new organisms. Both fields also consider the concept of modularity, where a discrete component is separated and studied from its surrounding environment, leading to a goal in synthetic biology, standardized biological parts. However, biological systems may also display principles of "emergence," where a system's properties may turn out to be greater than the sum of the properties of its individual components. This property may then complicate the synthetic design goals of creating systems by linking together individual parts. A fundamental question also remains regarding the extent to which biology, with its inherent complexity and "messiness," can be made into a fully quantitative field analogous to other branches of traditional engineering. Dr. Calvert stressed that both fields of systems and synthetic biology have become highly interdisciplinary and can draw on expertise outside of traditional life sciences departments. The presentation raised the question of whether new types of academic structures would be needed to house this type of cross-disciplinary research.

The new developments in these fields also raise interesting questions about data sharing and intellectual property. Electronic information, such as DNA sequences or computer code, is often the material being shared rather than physical samples. An "open source" ethic currently exists in some parts of the field, embodied by groups such as the BioBricks Registry of biological parts. Having such open source biological information available to the research community might speed developments in the field in the same way that an open-source computer code can speed computer software developments.

Another interesting question to consider is how easy synthetic biology currently is for nonexperts to perform. Despite the successes achieved in student competitions such as iGEM,[4] practical applications remain

[4] The 2008 International Genetically Engineered Machine (iGEM) competition Web site is available at: *http://2008.igem.org*. Accessed December 11, 2008.

some time away and, as the two previous presenters also highlighted, the inherent complexity of biological systems remains a great challenge. However, the synthetic biology community has taken several steps to openly discuss and write about potential risks that might be posed by technological developments in the field. Ethics-related sessions are included at the annual International Meeting on Synthetic Biology (SynBio), and the social science community has engaged the scientific community in considering the issues posed. In general synthetic biologists favor a self-governance model. However, such self-governance may not be as acceptable to all members of the NGO and public communities, some of whom have called for having a more inclusive public debate on the technologies and have pointed to a need to develop additional strategies to manage the potential risks that could arise from this technology.

Plenary 3: Introduction of the Breakout Sessions

After listening to the introductory panel survey several possible ways that the international community might think about life sciences and biosecurity issues, and also to the presentations highlighting scientific advances in emerging fields such as computational, systems and synthetic biology, the Forum participants considered the topics of the three working groups: (1) education and awareness, (2) oversight models, and (3) science advising. The chairs of each working group briefly summarized the objectives for their groups as well as some recent developments of relevance to their topics, so that all Forum participants would have a good sense of the workshop themes.

Leiv Sydnes (University of Bergen and past President of the International Union of Pure and Applied Chemistry [IUPAC]), Chair of working group 1 on education and awareness-raising, spoke on building a culture of responsibility. He highlighted some of the links between chemistry and biology and spoke of several ways in which the Chemical Weapons Convention (CWC) has brought chemical safety and security responsibilities into greater focus for practicing chemists. Industry initiatives including Responsible Care,[5] REACH,[6] and SAICM,[7] have also contributed to a greater emphasis on chemical safety and will lead to enhanced understanding of the toxicology of many chemicals being used. There has also been a greater focus on chemical safety and security as part of university

[5] More information is available at: *http://www.responsiblecare.org*. Accessed December 11, 2008.

[6] *Registration, Evaluation, Authorisation and Restriction of Chemical Substances*. Available at: *http://ec.europa.eu/environment/chemicals/reach/reach-intro.htm*. Accessed December 11, 2008.

[7] *Strategic Approach to International Chemicals Management*. Available at: *http://www.chem.unep.ch/saicm/*. Accessed December 11, 2008.

chemistry education than has been the case in biology. Dr. Sydnes spoke of the need for collective understanding and acceptance of the reasons for building a culture of responsibility among practicing scientists to make such a culture an integral part of each given discipline. He highlighted several features of an effective culture of responsibility, including: widespread acceptance of the scientific basis for professional responsibility; risk assessment as an integrated aspect of the profession; the inclusion of ethics; and continuous evaluation and adjustment as necessary. He concluded by suggesting several types of educational measures that might contribute to the development of cultures of responsibility, including greater focus on ethical considerations as part of school curricula at multiple levels, and greater incorporation of risk assessments into research projects as appropriate. In a similar manner, he suggested that greater awareness of the BWC and CWC and their implications might be useful tools to help educate both chemical and biological scientists.

David Franz (Midwest Research Institute), Chair of working group 2 on oversight models, spoke to the Forum next. Dr. Franz emphasized that the key challenge in considering standards and methods for research oversight is to protect scientific creativity and discovery, while simultaneously reducing the chances of the misuse of science to cause harm. He then explained the background of the creation and mission of the U.S. National Science Advisory Board for Biosecurity (NSABB), an advisory group created by the U.S. government and managed by the National Institutes of Health (NIH). The NSABB consists of 25 voting members appointed by the Secretary of Health and Human Services, and its charges include making recommendations to the U.S. government on criteria for identifying dual use research of concern (DURC), guidelines for oversight of dual use research, needs in biosecurity education, creation of a scientific code of conduct, policies governing publication, communication and dissemination of dual use research, and strategies for engaging the international community in a dialogue on dual use biology research.[8] The NSABB undertakes its mission through working groups on these various topics, and holds periodic public meetings to discuss the issues and progress. Dr. Franz presented the definitions of dual use research and dual use research of concern adopted by the NSABB, as well as highlights from the *Draft Proposed Framework for the Oversight of Dual Use Life Sciences Research: Strategies for Minimizing the Potential Misuse of Research Information* submitted by the NSABB to the U.S. government.[9] This document considers

[8] More information about the NSABB is available at: *http://oba.od.nih.gov/biosecurity/*. Accessed on December 13, 2008.

[9] Available at: *http://oba.od.nih.gov/biosecurity/pdf/Fframework%20for%20transmittal%200807_Sept07.pdf*. Accessed December 11, 2008. DURC is a more limited category than the NSABB's original charter, which was intended to cover general issues related to dual use research.

DURC to be only a small subset of dual use research. Oversight should focus on effective identification of such research followed by responsible conduct of research and dissemination of research results, not on prohibiting or restricting the research itself from being carried out. He raised several broad questions for working group 2 to consider, including what was needed versus what was being done, key international challenges, and areas of scientific consensus. He concluded with a suggestion that perhaps consensus could be found on the global nature of science, the rapid pace of scientific developments and the many benefits provided by these scientific advances, the need for a culture of responsibility and awareness, and the need for multiple approaches to address biosecurity and dual use issues.

Finally, Angelo Azzi (Tufts University and President of the International Union of Biochemistry and Molecular Biology [IUBMB]), Chair of working group 3, spoke about the roles of the scientific community in providing advice on biosecurity policy issues. Dr. Azzi explained that IUBMB, like IUPAC, was interested in codes of conduct. He suggested that life science organizations consider drafting a universal code of conduct as a unique document to be made available to everyone. He explained that IUBMB has used science as a vehicle to reach out to many countries including Iran, where IUBMB recently held a conference. IUBMB can also help contact and inform publishers and journal editors about these issues. Dr. Azzi also emphasized that it is important to present a clear case as to why the life sciences community is undertaking work on biosecurity issues. Bertrand Russell and Albert Einstein, for instance, were motivated to write their Manifesto in reaction against the proliferation of the hydrogen bomb. Dr. Azzi suggested that the community could create similar statements to better illustrate the level of danger from dual use biotechnology. He also suggested that realistic scenarios and better risk assessment tools could be helpful in presenting the problem. The concept of biosecurity can be used to move from a culture of fear to a culture of peace.

Plenary 4: Awareness About and Attitudes Toward Biosecurity

Plenary session 4 explored the results of several recent projects. Li Huang (Chinese Academy of Sciences [CAS]) discussed the history of biosecurity activities through the IAP, including the production and dissemination of the 2005 IAP Statement on Biosecurity. The IAP, then consisting of 93 academies of science throughout the world, formed a Biosecurity Working Group (BWG) in 2004 composed of the academies of science of China, Cuba, the Netherlands, Nigeria, the United States, and the United Kingdom. The working group drafted a biosecurity statement, which was launched in 2005, and has been endorsed by 69 of the IAP member academies (see Appendix D). It has also cosponsored several meetings including the first and second International Forums on Biosecurity, is planning to conduct biosecurity surveys in sub-Saharan Africa, and is developing an online biosecurity resource tool kit for member academies to help further their own national activities.

The BWG followed up the biosecurity statement launch by conducting two surveys of IAP member academies to examine ways in which academies have made use of the statement. The IAP statement consisted of a set of guiding principles that should be considered in developing biosecurity codes of conduct; and the results of the two surveys show that it has been translated into 8 languages, has been posted on numerous academies' Web sites, and presented to national authorities by 20 academies. Furthermore, seven academies have subsequently developed their own code of conduct and others have held conferences on topics related to biosecurity.

Dr. Huang also reported to the Forum several of the issues that the IAP BWG had encountered as it developed and disseminated the biosecurity statement. Some member academies felt that biosecurity as conceived in the IAP statement was not a high priority, or that natural biorisks were of far greater immediate concern than was laboratory biosecurity. Issues were also raised about risks from possible restrictions on sharing biological knowledge and information, and that such restrictions would be counterproductive to the goal of global biosecurity. In addition, there was concern about confusion over biosecurity terminology stemming from differing understandings and uses of the term.

Dr. Huang reported on recent initiatives from the CAS as an example of one academy that has undertaken additional biosecurity-related activities. The CAS has established biosafety committees and training programs at each of its life science institutes, has actively participated in international biosecurity discussions through groups such as the IAP and the WHO, and through two workshops: the CAS-COMEST symposia on ethics in science in Beijing and Shanghai in 2005,[10] and the upcoming international biosecurity workshop to be held in Beijing in late 2008.

The U.S. National Academies also has an active program of engagement in biosecurity activities. Recently, for example, the National Academies and the American Association for the Advancement of Science (AAAS) undertook a survey project on scientists' attitudes about biosecurity. Ronald Atlas (University of Louisville) served as the chair of the

[10] COMEST is the World Commission on the Ethics of Scientific Knowledge and Technology of the United Nations Educational, Scientific, and Cultural Organization (UNESCO). More information on CAS activities is available at: *http://english.cas.cn/*.

National Academies committee undertaking this work, and spoke to the Forum about the project.[11]

The study was undertaken to help address the lack of quantitative data on life scientists' attitudes toward biosecurity and dual use biology. By conducting surveys to gauge scientists' views on potential biosecurity risks and the roles that various groups should play, and then relating these to particular subpopulation demographics, the study committee hoped to enable the design of effective methods to engage these various groups of life scientists in biosecurity concerns.

The Web-based survey was conducted on a sample of 10,000 life scientist members of AAAS. Questions on the survey assessed respondents' perceptions of: the risk of bioterrorist acts, whether the respondent believed that the current research that he/she conducted was dual use, acceptance of options to address potential dual use issues, whose responsibility it should be to address such issues, and whether the respondent had personally taken any actions in response to concerns about dual use research. The 20 percent response rate (typical of Web-based surveys) limited the ability to generalize from the results. However, Dr. Atlas reported that the study committee was currently analyzing interesting trends in the data and looked forward to the public release. When finalized, the report will be made available on the National Academies Web site.[12]

Finally, Brian Rappert (University of Exeter) spoke about the project that he and Malcolm Dando (University of Bradford) had been undertaking along with additional international colleagues. The project explores the construction of effective biosecurity education methods, the purposes of such education, and how education might best engage its intended audiences. To help answer such questions, multiple seminars have been conducted in locations around the world. At the time of the presentation the group had conducted 26 seminars in life sciences departments in the United Kingdom and had conducted more than 70 seminars in the United States, South Africa, the Netherlands, Finland, Japan, Israel, India, Argentina, Uganda, Kenya, Ukraine, and Australia.

The seminars developed by the group seek to bring biosecurity discussions directly to researchers and students, and are usually held as part of regular university departmental seminars. They are also structured

[11] The Committee on Assessing Fundamental Attitudes of Life Scientists as a Basis for Biosecurity Education. More information is available at: *http://www8.nationalacademies.org/cp/projectview.aspx?key=48852*. Accessed December 11, 2008. The report of the survey results and analysis was still in progress at the time of the Forum, and official results could not be released to the group.

[12] Information about how to obtain the report, as well as information about other projects and events, is available on the National Academies Biosecurity Web site *http://www7.nationalacademies.org/biosecurity/*.

to engage seminar attendees and to foster conversation about topics on which there may not be consensus, such as the extent to which publication of dual use research should be restricted. Dr. Rappert reported on the broad results from the seminars. He indicated that, in general, participants felt that potential dual use experiments should or would be done, that the publication of research results should not be restricted, and that additional oversight was unlikely to be viable or desirable. He further reported that the interactive nature of the seminars demonstrated the importance of the process of active discussion and deliberation, as most participants initially felt that biosecurity was not an important issue, but became more engaged with the issue through participation. The results highlight the need for further education and awareness raising.

Finally, Dr. Rappert reported on continuing activities and initiatives in several of the countries visited as part of the seminar series, including the development of an educational module in South Africa and the implementation of additional biosecurity legislation in Australia. Looking to the future, he concluded by suggesting that further dual use education could serve different purposes in different contexts. In countries where biosecurity concern is currently high, such education might help support national calls to action. In countries in which there is some degree of awareness of biosecurity, it might help promote partnerships among countries and promote existing resources. Finally, in countries with no current interest in biosecurity, education could serve as the means to raise the issue and begin the process of engagement.

Plenary 5: The 2008 BWC Intersessional Meetings

Ambassador Georgi Avramchev (Permanent Mission of the Republic of Macedonia to the United Nations Office at Geneva and Chair of the 2008 Meetings of the Biological Weapons Convention) described his vision for the upcoming BWC meetings. He emphasized the importance that he placed on including the voices of the international scientific community in the discussions. The Ambassador summarized the BWC provisions and described the current intersessional process, which has proven to be a valuable mechanism to address technical topics agreed on by States Parties to be of particular importance. Although the intersessional meetings do not negotiate international treaty commitments, they serve to help bridge differences of opinion among member states by promoting common understanding, discussion, and an atmosphere of collaboration. The meetings have also proven to be valuable in broadening the participation and engagement of stakeholders beyond the diplomatic and security communities, and particularly expert communities in the life sciences, agriculture, public health, and education.

The Ambassador expressed his hope that the scientific community would provide valuable input into the topics to be considered over the next several years, while paying particular attention to the 2008 work program. The discussions at the August 2008 Meeting of Experts will consider: (a) national, regional, and international measures to improve biosafety and biosecurity; and (b) oversight, education, awareness raising, and adoption and development of codes of conduct. The December 2008 Meeting of States Parties will consider the results of the August technical discussions in the broader context of the BWC.

The Ambassador endorsed the goals of the Forum in encouraging communication and cooperation among international scientific experts and organizations and in putting the issue of biosecurity on such groups' agendas. He expressed his hope that this would raise awareness and strengthen the important engagement of scientific experts with the work of the BWC. The Ambassador also spoke about his plans for the upcoming meetings, and of several proposals to States Parties to make them even more effective. These included a proposed online resource for States Parties to share national approaches to dealing with biosecurity, biosafety, and oversight issues, in the hope that such an option might improve efficiency of information exchange among States Parties and increase available discussion time during the Meeting of Experts. Poster sessions and discussion panels of experts have also been considered, and the Ambassador looked forward to the additional side events that serve as opportunities for experts to interact and for stakeholder communities to inform delegations.

The Ambassador focused many of his remarks on concrete ways the scientific community might effectively contribute to the work of the BWC. In particular, he suggested holding events, both within the scientific community and as side events at the BWC meetings in order to: discuss biosafety, biosecurity, oversight and outreach; continue to produce reports and documents on such issues, which also served as valuable resources for the BWC Implementation Support Unit to draw on in preparing the background papers for the meeting; placing biosecurity issues on the agendas of scientific organizations; contacting national ministries of foreign affairs about meeting preparations; and participating in other ways, such as by serving on a BWC discussion panel or presenting a poster. The Ambassador also thanked the organizers and participants for the valuable contributions that their discussions at the Forum would make to his preparations for the upcoming 2008 BWC meetings.

SUMMARY OF BREAKOUT SESSIONS

Summary of Working Group 1: Education and Awareness Raising

Chair: Leiv Sydnes
Rapporteur: Alastair Hay
Summary prepared by Jo Husbands

Background

Working group 1 discussed how to improve awareness among scientists about issues related to the use of the life sciences by states or terrorists for biological weapons. The group's suggestions are intended to help foster and sustain a culture of responsibility within the scientific community about the risks of misuse and the roles that scientists can play to help reduce them. To provide background for the discussions, several individuals were invited to make presentations about their activities.

Working Group Presentations

Dual Use Seminars. A series of presentations highlighted the lessons learned by collaborators in the course of an international project, "The Life Sciences, Biosecurity, and Dual Use Research," organized by Brian Rappert (University of Exeter) and Malcolm Dando (University of Bradford) (see Plenary 4 for more information). The project, sponsored by the Alfred P. Sloan Foundation, has conducted more than 70 seminars with over 1,600 total participants in a dozen countries during 2006 and 2007. Several of the project's international partners made presentations about their experiences with those seminars, as well as broader biosecurity issues in their countries.

Katsuhisa Furukawa (Research Institute of Science and Technology for Society) began with an overview of Japan's experience, the "taboo" on biological weapons because of Japan's use of them in World War II, and the role of biological scientists in Aum Shinrikyo. In Japan, biosecurity is a relatively new issue and is not considered to be a major risk relative to other security threats; therefore, activities there are focused largely on raising awareness of the issue. A recent law has substantially increased the requirements for security at Japanese laboratories. The remaining challenges are:

- How should the knowledge and expertise associated with dual use research be managed?
- What responsible management structure should be instituted at universities and academic institutions?

A continuing problem, however, is that the focus remains more on agents and equipment rather than on knowledge and techniques.

Dr. Furukawa provided additional details about Japanese government policies and actions. With regards to engaging scientists and officials, he found substantial reluctance on their part to consider the possible negative sides of advanced research. It was effective, however, to use examples of advanced research experiments to raise dual use concerns. He found that visits by experts and scholars, as well as general interaction with the international scientific community, were helpful in raising interest. Dr. Furukawa also identified several additional steps that could be taken:

- The exchange of experience and information about effective guidelines to identify dual use experiments of concern.
- The creation of an educational module by gathering specific case studies of the misuse of scientific research to inform science students and researchers about the dual use challenges. It is also desirable to use such educational modules to educate other stakeholders, such as managers and administrators in universities, research institutions, and companies, as well as stakeholders in the government and media, when appropriate.
- The exchange of information about the efforts to address the challenge associated with the access to research programs at universities by those individuals about whom there is potential concern for misuse, including those foreign students from countries that pose proliferation concerns.
- A more coordinated Asian region approach to assisting other countries' efforts on biosecurity.

Animesh Roul (Society for the Study of Peace and Conflict) commented that, like Japan, bioterrorism risks in India were considered less urgent threats than terrorism involving nuclear and chemical materials and facilities. Biosecurity is also very much concerned with risks to agriculture, and most of the potentially relevant regulation in India relates to pests or diseases that threaten crops or livestock.

Through the seminars that he had helped organize in several universities and research institutes, Mr. Roul concluded that most Indian biological scientists were: (1) very confident about their work ethics and responsibilities; (2) quite averse to the idea that there was something that they needed to learn and absorb, especially since it might restrict their research; (3) skeptical about possible misconduct by their own colleagues and scientists generally; and (4) convinced that the bioterrorism/biodefense issue was basically a Western (particularly U.S.-generated) concept and phobia.

Mr. Roul found that, although the level of acceptance of dual use concerns was low, there was a prospect for progress on increased awareness of this issue in the future. Senior/retired scientists were generally more interested and accepting of the dual use threat issue. Interactive seminars are always helpful in getting answers and ideas, but to develop that awareness further, focused discussions and workshops could be used to engage scientists. He concluded that a great deal of work remained toward raising the awareness level within the scientific community.

Chandré Gould (Institute for Security Studies [ISS]) described the biosecurity efforts in South Africa, as well as the lessons learned from seminars and meetings that she had helped organize in Kenya and Uganda. In the South African case, she noted that, although the government had put national measures in place against biological weapons and was actively engaged on the international level, bioterrorism was not regarded as a significant risk. There was also limited outreach capacity and little engagement with the scientific community. She also described a variety of activities and involvement by parts of civil society, although engagement was low relative to other biotechnology issues, such as concerns about genetically modified organisms (GMOs).

Dr. Gould helped organize several seminars for Rappert and Dando, and found that previous contacts with the universities made the task easier, although she nonetheless found it difficult to convince several universities and departments of the importance and relevance of dual use issues. The involvement of academics from outside the country made the topic more attractive for South Africa, although slow responses complicated logistical arrangements. She reported that the seminars had evoked a mixed response from institutions in terms of research oversight and policy responses to biosecurity issues. The most positive response came from the National Institute for Communicable Diseases. This was the third of several meetings and led Dr. Gould to conclude that repeated interactions built trust, interest and buy-in.

In Kenya and Uganda, Dr. Gould and her colleagues found that "biosecurity" was a term associated with GMOs, which complicated discussions. Again, she found that logistical issues presented significant challenges; the presence of an ISS office in Nairobi was essential, as well as local assistance and buy-in. Here the contacts that Rappert and Dando had made with a leading Ugandan scientist at the 2006 Royal Society workshop on trends in life sciences that were relevant to the BWC, proved to be extremely helpful. It proved difficult to convince several universities and departments of the importance and relevance of dual use issues, and there was also sensitivity from some about becoming involved in a national security issue (as was also true in South Africa).

In spite of these obstacles, Rappert and Gould were able to conduct a series of very effective meetings involving policy makers and civil society representatives. They were also able to use the time in Kenya and Uganda to develop additional contacts, including Ugandan parliamentarians. These discussions helped lay the groundwork for a workshop on biosafety and biosecurity organized by the Uganda National Academy of Sciences (see working group 3), which brought together scientists and policy makers from eastern and southern Africa, and stimulated further collaboration among a number of organizations in the region.

Recently, Dr. Gould has been engaged in developing an educational module on dual use issues relevant to specific national circumstances, including national regulatory environments. In the future, she will continue collaborating on policy discussions and development in other countries with the contacts that she has developed. Given the reluctance of many scientists to become engaged in dual use issues, she emphasized the importance of repeat visits and relationship development. She noted the frequent difficulty of identifying the right people to talk to on the first visit. Dr. Gould concluded by saying that international policy development is a slow and sometimes painful process.

Lessons from Biological Weapons Programs for Education and Awareness Raising. Iris Hunger (University of Hamburg) started by discussing the current focus on scientists as an important target for biological weapons control efforts. This is illustrated by the focus of the 2008 BWC intersessional meetings, cooperative threat reduction programs directed at former weapons scientists, and various national measures to restrict scientific activities, such as increased controls on dangerous pathogens. She drew on a number of case studies to ask whether this focus was justified. She asked if it was indeed true that scientists had a decisive influence on the initiation, shape, and elimination of bioweapons efforts?

Among the possible "proactive" activities of scientists, based on an analysis of several historical cases, she cited: lobbying for the establishment—or for the termination—of a biological weapons program; the unrequested development of proposals for enhanced or new types of biological weapons; and the conscious distortion of technical and scientific data to hasten or hinder a weapons program.

Dr. Hunger's research found that there were cases (South Africa and Japan) where a single scientist or physician essentially started and ran a biological weapons program. There was also a case (Germany) where scientists did lobby for a weapons program, but were largely unsuccessful. She found that often scientific advisory bodies worked on the basis of responding to questions, instead of setting an independent policy agenda. There were also several instances (the United States and the United Kingdom) of scientific advice being ignored, and scientists complaining about not being taken seriously. She cited a case (the Soviet Union) where scientists successfully proposed and pushed for new types of weapons. She also cited a case (the United States) where a group of scientists was instrumental in ending a weapons program. Her research has led her to three hypotheses:

Hypothesis 1: (a) Scientist-based approaches are most promising if aimed at identifying biological weapons programs ("whistle blowing") and slowing and stopping them. (b) Scientist-based approaches are less promising if aimed at preventing weapons activities; that is, stopping a country or a nonstate actor from developing weapons.

Hypothesis 2: (a) Scientists contribute to bioweapons efforts through (i) conscious participation, (ii) negligence ("I knew, but I did not care"), and (iii) willful ignorance ("I did not know"). (b) Scientist-based approaches are most promising if aimed at preventing scientists from becoming bioweaponeers, because of a lack of knowledge.

Hypothesis 3: Scientist-based approaches are most promising if aimed at democratic societies and/or open scientific communities.

These hypotheses have implications for any program that aims to raise awareness or provide education about dual use issues.

An Example of Education Modules: IUPAC and the Organization for the Prohibition of Chemical Weapons (OPCW). Alastair Hay (University of Leeds) presented an example of an educational module developed to address dual use issues for undergraduates in chemistry. The module is a joint effort between IUPAC and OPCW. The module is intended to raise awareness about the Chemical Weapons Convention. Dr. Hay cited some of the following challenges to the education and outreach efforts related to the CWC:

- Relevance and ownership by teachers and students in many countries—"the CWC is 'someone else's problem' ";
- Concerns about the negative impact on the public image of chemistry;
- Limited knowledge of the CWC and dual use issues among chemistry teachers at all levels;
- Little attention to ethical issues of any sort in the curriculum;
- Remoteness of the CWC structure to the educational system.

The approach that the IUPAC education group has taken is to place chemical and biological weapons in the larger context of multiuse chemicals. The developers of the educational module consider it essential to start with the beneficial aspects of multiuse chemicals and then move on to the issues of abuse and misuse. The module is targeted at chemists and chemistry educators in the domain of influence of IUPAC and OPCW. The project considered it important to pilot materials with educators and to evaluate them from the beginning, in order to refine materials and approaches. The materials are designed to be delivered over the Web and were piloted in several countries to address language issues. The project also sought to enlist partners for broad dissemination.

The module moves from the beneficial effects of natural and manufactured chemicals to familiar examples of misuse such as ephedra and methamphetamine. It emphasizes that the choices about beneficial use, misuse, or abuse lie in our own hands. The role for science education is thus to consider issues of access to information and the risks of diversion of readily available materials. This leads to questions of who has responsibility with the aim of fostering understanding and ownership of ethical responsibility. The module then turns to chemical weapons, again emphasizing the multiple uses of the basic chemicals, and providing examples of both historical (World War I) and more recent use (civilian areas in Iran and Iraq). The project is now completed and the material—text and pictures—is available on the Web along with four background papers in six languages.[13]

Building a Culture of Responsibility. Gerald Epstein (Center for Strategic and International Studies [CSIS]) gave a presentation with two goals: (1) to recap the development to date of a "culture of responsibility" or governance within the synthetic genomics community; and (2) to describe a new CSIS project to broaden the culture of responsibility beyond synthetic genomics, and to extend it outside the scientific community.

Synthetic genomics is the ability to construct and "boot" long strands of genomic material, and thereby to construct organisms (viruses and eventually higher life forms) within specified genomes. He reminded the group that synthesis is not the only way to construct genome-length strands of DNA, and that synthesis technologies are pervasive in biological applications other than constructing genomes. In terms of the attitudes of the synthetic genomics community toward governance, Dr. Epstein commented that many leading synthetic genomics researchers have initiated and/or participated in governance activities. Some researchers, however, have a problem with "arbitrary" focus on one way to construct genomes while ignoring others, and some worry about catering to what they feel to be unwarranted public perceptions.

[13] Web site for "Raising Awareness: Multiple uses for chemicals and the chemical weapons convention (IUPAC Project 2005-029-1-050)" is available at: *http://www.multiple.kcvs.ca/*. Accessed on December 11, 2008.

Dr. Epstein reviewed a variety of governance proposals and studies, both from inside and outside of government, though to date most of the work has been on the outside. He asked whether there is a "window for governance" in the development of a technology where effective governance mechanisms might be desirable. This window would occur somewhere between when the technology is nascent and controls are infeasible or unnecessary ("too early to tell"), and when it has matured and become so pervasive that control is impossible ("too late to change").

Dr. Epstein commented that, next to nanotechnology, synthetic genomics may be the most assessed and analyzed field that does not yet exist. It is already impossible to have an internal, scientific community-only discussion of potential governance mechanisms. The press and key stakeholders are watching very closely, but the problem of defining who is a "stakeholder" is a major issue. This is relevant to broader aspects of creating a culture of responsibility for the life sciences, which must also address issues of awareness and education within the community, engagement with other communities and governments involved in managing relevant risks, engagement with other stakeholders and the public, and participation in global governance.

Dr. Epstein emphasized the many facets of the deliberate use of biology to cause harm and the many communities that would have to be engaged in preventing, detecting, and responding to incidents. Each community must see how its activities play a role in reducing biothreats; each community must understand how biothreat reduction activities could affect its own mission, if at all; and each community needs to know about other communities, and how their actions impact on one another. Yet none of the communities has the reduction of biothreats as its primary mission.

Traditional top down, hierarchical governance structures are poorly suited for issues such as biorisk management, which Dr. Epstein argued are highly decentralized, highly interdisciplinary and cross-community, rapidly evolving and highly S&T dependent, significantly driven by non-state actors, global more than international, and which lack consensus as to the nature and magnitude of the problem. The new CSIS project, the Global Forum on Biorisks, is intended to address this problem. The forum is based on the belief that, whatever the answers are, they will arise from a bottom-up, decentralized process of engagement, interaction, assessment, and analysis among all relevant professional communities around the world. The project is implementing a highly interactive, professional community-based Web portal to facilitate these interactions. Dr. Epstein believes that it will be an ideal environment in which to continue the kinds of discussions taking place in the forum, and invited the participants to join when the forum is up and running later in the year.

Working Group Discussions

The working group discussed questions and issues raised by the presentations, as well as reflected on several questions that were posed to the group to help initiate dialogue. The questions were:

- What kinds of awareness raising and educational activities are needed? What are some examples of current projects and activities?
- What particular challenges are faced by those trying to develop a "culture of responsibility"?
- What are some of the lessons learned from your work and what advice would you give to others planning activities?
- Are there particular unmet needs or opportunities on which groups might focus?

Results

The group made two general suggestions. First, along with the other working groups, group 1 noted the difficulties posed by the many different meanings of the term "biosecurity." This can lead to substantial confusion and the group suggested, therefore, that when biosecurity was discussed, the issues should be presented in simple, easily understood terms. For example, biosecurity could be broadly defined as "measures to reduce the risk from the natural, accidental, and deliberate spread of disease."

Second, the group supported, to varying degrees, ongoing activities to develop and promulgate codes in the life sciences. The members of the group did not agree about how much codes of conduct would contribute to efforts to *prevent* misuse in this area, given historical experience. They did agree that codes of conduct as part of a broader approach to biosecurity, could help both to raise awareness and to foster a culture of responsibility in the scientific community, and thus could contribute to educational efforts. There was strong support for the IAP Statement on Biosecurity (see Appendix D) as providing essential principles that any code of conduct should include. The group, therefore, encouraged governments to support initiatives to implement the Statement through the development of new codes by national scientific bodies, such as academies and professional societies, or the modification of existing codes to include biosecurity issues. There was also discussion about the importance of encouraging participation by as many stakeholders as possible in the process of drafting codes of conduct, so that, through discussion, they will share and enhance awareness of the issues. In addition, the view was also expressed that each stakeholder institution should be encouraged to develop its own codes, applicable to its own circumstances, and articulated to its own audiences.

The group also made suggestions for actions in four specific areas:

Awareness Raising. The group agreed that it was vital to raise awareness among scientists, and life scientists in particular, about: (a) the threats from natural, accidental, and deliberate spread of disease; (b) the history of biological warfare and biological weapons programs; (c) the dual use dilemma posed by life sciences research; (d) the prohibitions and obligations imposed by the BWC; and (e) national laws and regulations intended to mitigate the risk that life sciences might be misused. There was also discussion that such awareness-raising efforts should not necessarily be limited to scientists alone, but should include other stakeholders, such as managers and administrators in universities, research institutions, and companies, as well as stakeholders in the government and media.

The group suggested that a program to raise awareness should be developed in such a way that it could benefit from and support the BWC process. Because awareness raising is a continuing process that will have to be sustained over many years, it is important to involve governments, since they can provide resources and support, even if the efforts are carried out by independent scientific bodies. An endorsement by the BWC process could help commit member states to this effort.

Education. The Group made a number of suggestions related to education. First, the scientific community needs to ensure that ethics training was mandatory and is supported by adequate resources. Good teaching material with appropriate case studies will be needed to support this training. The materials will need to include the issues identified as important under "awareness raising," as well as materials related to a scientist's personal responsibility for the conduct of his or her own research or research that she/he supervises. Although some materials exist for these purposes, education packages need to be developed, and the Group suggested that strong support, including essential financial resources, be dedicated to these endeavors.

General ethics training is already provided for scientists by many universities and some of this material may be useful for education. As education efforts continue and new programs are developed and implemented, there will be a need to share best practices. The Group therefore suggested that a clearinghouse or repository be established to identify useful educational resources and to share best practices and lessons learned from different training experiences.

Industry and the Private Sector. The group concluded that there is a need to expand the engagement of industry and the private sector in the awareness-raising process. Many of these organizations have experience with training and can contribute to education. The group suggested that, where possible, CEOs and senior scientists in industry be approached and encouraged to become involved.

Resources. In 1986, the 2nd BWC Review Conference identified the importance of education and awareness for the successful implementation of the Convention's goals. Current evidence suggests that awareness of the BWC and the challenges of biosecurity remain low within the life sciences community. As mentioned above, raising awareness will take considerable time and resources. In particular, funding will be needed to design appropriate teaching materials and to assess their efficacy. The group suggested that the States Parties to the BWC should provide sustained funding for education modules and awareness raising, and that these Parties should commit at the 7th Review Conference in 2011, to report annually on their efforts to promote education.

Finally, the group suggested that a task force be established under the auspices of the IAP Biosecurity Working Group to consider:

- Where and how best to establish and operate a clearinghouse;
- How to achieve the objectives for education, for awareness raising and for involving industry and the private sector; and
- How to secure resources to fund the various initiatives.

Highlights of the group discussions and suggestions were presented by Alastair Hay to the entire Forum in one of the final plenary sessions.

Summary of Working Group 2: Oversight of Research

Chair: David Franz
Rapporteur: Neil Davison
Summary prepared by Ben Rusek

Background

During the proceedings of working group 2, Chair David Franz guided the group on a discussion of important issues related to research oversight. Several individuals were invited to make presentations about their activities, so as to provide background for the discussions.

A Prototype Protective Oversight System. Elisa Harris (Center for International and Security Studies at the University of Maryland [CISSM]) presented her project's proposed system for oversight of dual use research. She explained that increasing attention to the dual use problem has raised questions about the adequacy of existing oversight. Concerns about new dangers can arise from unexpected results, from misuse of legitimate research, and from the blurred line existing between offensive and defensive research. Recent examples include experiments involving mousepox, polio virus, and influenza.[14] Ms. Harris said that any measures taken to address dual use issues must be balanced between protecting the right of scientific investigation and enforcing the norm against the destructive applications of biology. It is also important to reassure scientists that they would not be subject to excessive regulation, and to reassure society that the power of biology was being used appropriately.

As Ms. Harris explained, the essential features of the CISSM model are: it is narrowly focused and excludes most biomedical/pathogen research; it is readily implemented and the definitions of covered activities are provided in checklist form; it is a practical response to the threat and combines agent- and activity-based approaches; and it is a tiered design with local review being most predominant. Licensing or registration of certain personnel and facilities and independent peer review of certain research projects are key elements. The proposed system would apply to all relevant institutions: government, academic, and industry. CISSM also proposes oversight methods on the national and international level. National review bodies would oversee and approve research of moderate concern (e.g., work with specific listed agents, particularly activities that enhance virulence, transmissibility, or weaponization). The proposed global implementing body would oversee and approve research of extreme concern (e.g., work involving the most dangerous of currently known pathogens, or possibly resulting in the creation of a significantly more dangerous pathogen).

Ms. Harris recognized that the international oversight arrangements were not going to happen overnight. She discussed some of the incremental steps that would need to take place in order to move toward the CISSM model. These included implementing codes of conduct and education and

[14] Jackson, R.J., A.J. Ramsay, C.D. Christensen, S. Beaton, D.F. Hall, and I.A. Ramshaw. 2001. Expression of mouse interleukin-4 by a recombinant ectromelia virus suppresses cytolytic lymphocyte responses and overcomes genetic resistance to mousepox. *Journal of Virology*, 7(3):1205-1210; Wimmer, E. 2006. The test-tube synthesis of a chemical called poliovirus. *EMBO Reports* 7(Special Issue):S3-S9; Tumpey, T. M., C.F. Basler, P.V. Aguilar, H. Zeng, A. Solórzano, D.E. Swayne, N.J. Cox, J.M. Katz, J.K. Taubenberger, P. Palese, and A. García-Sastre. 2005. Characterization of the reconstructed 1918 Spanish influenza pandemic virus. *Science* 310(5745):77-80.

training programs, including dual use review requirements in national biological safety programs, harmonizing national laws and regulations, building on existing WHO guidelines for lab biosafety and biosecurity, and developing dual use guidelines for member states.

Israeli Perspective on Biotechnological Research Oversight. David Friedman (Institute for National Security Studies, Tel-Aviv University and the Israel Academy of Sciences and Humanities) explained that Israel is in the top 10 countries in the life sciences and in the top three in some life science fields, and that Israeli scientists are very concerned about biosecurity issues. He explained that a steering committee on issues in biotechnology in the age of terrorism was established in Israel to address the problem of biosecurity threats. Its members were appointed jointly by the president of the Israel Academy of Sciences and Humanities and the head of the Israel National Security Council (NSC). The report resulted in changes to Israel's existing legislative infrastructure and made eight key recommendations:

1. Publicize the ongoing effort to raise awareness and understanding of the risks associated with the biological threat in general, and with dual use biological research in particular, among Israel's life and medical science community.
2. Implement legislation designed to prevent the seepage of organisms, material and information to potential terrorist elements and formulate specific long-term comprehensive biosecurity legislation.
3. Adapt existing biosafety oversight procedures to also ensure biosecurity and delegate responsibility for the enforcement of biosecurity to existing institutional biosafety committees (renamed "biosafety and biosecurity committees") in the academic sector.
4. Create an itemized core list of dangerous agents (adopted from the U.S. Department of Health and Human Services select agents list). The list should be reviewed and updated annually, as required.
5. Establish a system to oversee and approve dual use research projects by an internal mechanism based on the judgment of the academic community.
6. The Israel Science Foundation and government research foundations must require, as part of their approval process, biosecurity approval from the institution in which the research will be conducted.
7. Establish a system to oversee the Israeli import of dual use biological laboratory equipment and biological agents, as well as the sale of these items in the local market.
8. Establish a biosecurity regime or National Biosecurity Council under the Ministry of Health (MOH).

Dr. Friedman explained that the committee's report and recommendations were approved by the NSC and by the Israel Academy of Sciences and Humanities. A deliberation is taking place in the MOH regarding the implementation of the recommendations. The Committee for Science and Technology of the Israeli parliament is discussing enacting a law regarding research with pathogenic strains based on the committee's recommendations. Dr. Friedman reported that he was also taking part in preliminary deliberations to establish awareness, consciousness and education programs (workshops, symposia, etc.) for the life sciences community.

Dual Use Education and Review Within a U.S. University Consortium. Ruth Berkelman (Emory University) discussed efforts on research oversight undertaken by the Southeastern Regional Center of Excellence for Emerging Infectious Diseases and Biodefense (SERCEB), which is one of the 10 regional centers of excellence for biodefense policy sponsored by the U.S. NIH. SERCEB consists of six primary universities: Duke University, the University of North Carolina, Emory University, Vanderbilt University, the University of Alabama, and the University of Florida. The steering committee is composed of one researcher from each of these schools (plus the University of Michigan).

SERCEB conducts relevant biosecurity activities under its Policy, Ethics, and Law (PEL) core theme.[15] It includes education and awareness raising (biosecurity, dual use, biosafety), dual use review, science in the event of an emergency, emerging infections, and global health and policy engagement. The key oversight effort under this theme is the SERCEB PEL Dual Use Educational Module. The module was developed as a tool to teach scientists (senior scientists, students, and laboratory technicians) about biosecurity and the dual use dilemma. It walks the user through a scenario of a Ph.D. candidate facing dual use concerns in thesis work and provides the user with background on legal, ethical and policy implications. It was launched in 2005, has had more than 650 users to date, and is currently under additional revision. Three of the six universities have launched dual use modules and the others are examining the issue.

In addition, a dual use review of all SERCEB-funded projects takes place through the SERCEB Steering Committee. The steps in the review process are:

1. The steering committee receives research proposals for funding, flagging those proposals that could have dual use potential.
2. The committee sends proposals to the PEL core; PEL core members

[15] Available at: *http://www.serceb.org/pel.* Accessed December 11, 2008.

review the proposals individually (considering experiments of concern and NSABB criteria), before sharing thoughts collectively.

3. The committee receives PEL suggestions and/or follow-up questions and sends them to investigators.

4. The committee urges investigators to address PEL's concerns before disbursing funds.

Dr. Berkelman concluded by stating that despite national guidelines or laws governing dual use research, it will still be important to continue to educate scientists on the topic.

Addressing Risks of Research Misuse: A Funder's Perspective. David Carr (The Wellcome Trust) discussed research oversight from the perspective of a major biological science funding body. First he presented several reasons for the Wellcome Trust's attention to biosecurity. He listed new legislation on antiterrorism and export control, ongoing parliamentary attention, the involvement of UK Foreign and Commonwealth Office staff with the BWC codes of conduct discussions, and other international developments. The 2001 terrorist attacks and the Fink report and subsequent U.S. policy were also important developments.[16]

Mr. Carr explained that the Wellcome Trust released a position statement on bioterrorism and biomedical research in November 2003.[17] He favors a system of self-governance in the scientific community. Self-governance would be the most effective, appropriate and sensitive system and would best reduce the risk of misuse, without imposing onerous regulation. The Wellcome Trust system employs the strong existing funding framework based on peer review, where the host institution is responsible for ensuring that the requirements of all regulatory authorities are met. For the rare cases in which additional ethical and social issues are raised, advisory mechanisms are in place.

In 2005, the three largest funders of life sciences research in the United Kingdom—the Biotechnology and Biological Sciences Research Council (BBSRC), the Medical Research Council (MRC), and the Wellcome Trust—signed a follow-on policy statement.[18] It made changes to the procedures in the 2003 key statement in four areas. It added: (1) guidance for

[16] The Fink report, *Biotechnology Research in an Age of Terrorism*, takes its name from the committee's chair, Gerald Fink of MIT. (National Research Council. 2004a. Washington, DC: The National Academies Press).

[17] The statement is available at: *http://www.wellcome.ac.uk/About-us/Policy/Policy-and-position-statements/WTD002767.htm*. Accessed on December 11, 2008.

[18] The statement, Managing Risks of Misuse Associated with Grant Funding Activities, is available at: *http://www.wellcome.ac.uk/About-us/Policy/Policy-and-position-statesments/WTX026594.htm*. Accessed December 11, 2008.

applicants through the introduction of an added question on application forms; (2) guidance for referees with explicit mention of research misuse as an issue to consider; (3) guidance for funding committees, including a process for assessing cases where concerns were raised; and (4) modification of good practice guidelines to include specific references to research misuse. Mr. Carr explained that the policy has been in place for three years with some evidence that it has increased the number of people who were considering the risks of research. The Wellcome Trust has received a range of responses to the application-form questions, and he saw that as evidence that the policy was encouraging at least some applicants to consider if risks were associated with their proposals. So far a small handful of cases have been flagged for further consideration internally, but none have raised risk-benefit concerns that have impacted funding decisions. The Wellcome Trust has not formally assessed the impact of the statement on awareness raising, but hopes that it has contributed. Mr. Carr concluded by saying that dual use risks need to be considered by scientists at all levels of the research process.

Working Group Discussions

After the formal presentations, David Franz led the working group through a discussion of biosecurity definitions, elements of biological research oversight that were under way and that were proposed, the best principles for research oversight, international challenges to oversight, and possible next steps for the oversight of dual use research. The group discussed several research oversight methods that could be employed by interested groups.

The group agreed that any oversight mechanism must not unduly limit scientific research and scientific progress and noted that the majority of research in the life sciences would not fall into the dual use category. The aim should be to focus on research with the highest potential for risk, such as the seven experiments of concern cited in the Fink report.[19] The key is to reduce risk—including risk from unanticipated results—but not to stifle research. This can only be done in a dynamic fashion through regular re-evaluation of current activities and evaluation of new science and technology developments. Novel education and awareness raising methods (as discussed in working group 1) were important to underpin any potential oversight measures. A "toolkit" of different types of measures could be designed to help governments, organizations and

[19] NRC (National Research Council). 2004a. *Biotechnology Research in an Age of Terrorism.* Washington, DC: The National Academies Press, p.12.

individuals accomplish this. Oversight measures should coincide with a certain amount of self-governance and self regulation.

The elements of an oversight toolkit could include methods for preliminary research review, review before dissemination and/or communication of results of proposed research of concern, pathogen controls, personnel and facility accreditation, and limits on technology (knowledge, materials, equipment) transfer.

The group reviewed several oversight mechanisms proposed or already under way around the world. For example, the group discussed the statement released by journal editors in 2003,[20] and heard insights from the forthcoming NA/AAAS survey. Dr. Franz led the discussion on the oversight methods proposed by the NSABB, of which he is a member, and mentioned that the NSABB oversight framework was awaiting a U.S. government response/decision.[21]

The group saw some common elements among the methods presented. All showed a strong influence from the Fink report recommendations. The oversight methodologies focused on the areas of highest potential risk (although disagreements about risk categorizations remained). Each recognized the importance of education and awareness raising and all of the systems built on existing ethical and biosafety policies and procedures already in place.

The group also identified some significant differences in the approaches. Some emphasized self-governance, and this was evident in the systems employed by the Wellcome Trust, BBSRC, MRC and SERCEB. The NSABB, CISSM, and the Israel Academy/NSC proposal tended to make binding regulations the primary barrier to misuse. Several participants focused on the mechanism of reviewing research before the dissemination and communication of the research results/data, as an additional layer of protection to capture unexpected dual use results that might be dangerous.

Results

Principles for Research Oversight. After addressing existing mechanisms, the group discussed some broad principles important for research oversight. Again members emphasized that it was most critical to balance the risk of stifling research with the benefit of increased scrutiny of research. This

---

[20] Journal Editors and Authors Group. 2003b. Statement on the consideration of biodefense and biosecurity. *Nature* 421:771.

[21] More information is available at: *http://www.oba.od.nih.gov/biosecurity.* The oversight framework is available at: *http://www.oba.od.nih.gov/biosecurity/pdf/Framework%20for%20transmittal%200807_Sept07.pdf.* Accessed December 11, 2008.

balance could be tilted positively by involving scientists in the process and raising awareness of the dual use research risk. Since 21st century biological science is truly a global undertaking and pathogens do not respect international boundaries, many participants noted that oversight needed to take a more global perspective. Some degree of non-discriminatory harmonization and integration in conformity with international and/or regional guidelines, and in conjunction with national oversight, would help this effort.

A toolkit of measures for countries to draw upon could be constructed as follows: At the individual researcher level, basic awareness of the problem is critical, and voluntary oversight linked to funding could help educate. At the institutional level the focal point would be peer review and oversight. The national focal point would be oversight. The regional and international focal point would be guidelines.

International Challenges to Oversight. The group identified several key unanswered questions and international challenges.

1. *Institutional arrangements*: What role should international bodies (e.g., WHO guidelines) play? Should responsibility for oversight at the international level be given to existing institutions, or should new institutions be created?
2. *Capacity building*: How can one ensure support for research, disease surveillance and public health and assistance to countries to implement dual use research oversight measures?
3. *Harmonization across borders*: How can a level playing field be assured when the choice of elements in the biosecurity toolkit may vary from country to country?
4. *Risk assessment and prioritization*: How does dual use risk measure against other concerns, for example biosafety and endemic disease?
5. *Globalization and market forces*: How can scientists be prevented from avoiding research in dual use areas because of perceived overregulation, and be dissuaded from outsourcing to escape regulations?

Other questions raised by the group include: Do some oversight measures miss certain sectors such as industry, public health, and government (including military research)? How does one put a perceived legitimate oversight system in place for biodefense research (where the majority of dual use research takes place), and ensure secrecy as well as transparency? Is it possible to monitor and evaluate (or ensure compliance and enforce) oversight? What are the mechanisms for follow-up? By whom? What are the consequences for violation/noncompliance? How does one treat whistle blowing or early "intervention" and focus on awareness among researchers and adherence to existing rules, while avoiding a culture of suspicion? What kind of publication/dissemination guidelines for review and dissemination are needed at the individual and institutional levels prior to publication review?

The group agreed that research oversight, if done correctly, could have positive effects. Transparent oversight can improve societal support for responsible research. Self-governance steps already taken by scientists can preempt top-down overregulation. Good biosecurity can be good for business. The group suggested that scientists are used to regulation, but that they have an aversion to additional regulation perceived to be unreasonable; scientists are open to reasonable and feasible regulation, but only in an area where a tangible risk exists.

Suggestions for Next Steps in Oversight. David Franz suggested that incremental steps are important and that good suggestions had been made during the discussions that could help in the process of establishing oversight measures. The discussions also highlighted a number of common measures found in the various proposals. Moreover, those implementing regulatory systems should also remember that "perfect is the enemy of the good."

International organizations such as the WHO, UN Food and Agriculture Organization (FAO), the World Organization for Animal Health (OIE), OECD, and IAP have the convening power to bring together stakeholders and they can make an important contribution. The group suggested that the IAP produce a statement on research oversight as a follow-up to the 2005 IAP biosecurity statement. In addition, organizations could build dual use review into existing biological safety mechanisms, building on quality guidelines like Good Laboratory Practice (GLP), Good Clinical Practice (GCP) and Good Management Practice (GMP) to move toward "Good Biosecurity Practice" (GBP). Research oversight would benefit from adapting existing review mechanisms such as institutional biosafety committees that handle research involving recombinant DNA. The group suggested involving existing and relevant stakeholder organizations, for example the European Biosafety Association and the American Biosafety Association and others. The group also suggested some novel alternative methods and tools to link awareness raising and oversight, such as: an online advice portal for scientists on how to handle potential dual concerns in their research; improvements to OECD's biosecurity codes website;[22] a biosecurity "Wikipedia;" biosecurity posters for laboratories and special pages for journals that present the issue; and a systematic

[22] Available at: *http://www.biosecuritycodes.org*. Accessed December 11, 2008.

evaluation of the effectiveness of existing approaches to create more lessons learned and best practices.

Highlights of the group discussions and suggestions were presented by Neil Davison to the entire Forum in a plenary session.

Summary of Working Group 3: Science Advising

Chair: Angelo Azzi
Rapporteur: Ralf Trapp
Summary prepared by Katherine Bowman and Ralf Trapp

Background

Working group 3 focused its discussions on the role of the international community in providing scientific advice on issues related to biosecurity. To facilitate discussions among participants, several of the working group sessions incorporated brief presentations. The first talks were designed to highlight a few of the international governmental organizations that might serve as potential venues for addressing biosecurity-related issues and that might be valuable partners for the scientific community in considering these topics. Additional presentations highlighted recent biological and chemical security activities that had been undertaken by national academies of science and scientific unions in partnership with national and international organizations.

UNESCO. Lucy Hoareau (Division of Basic and Engineering Sciences, UNESCO) provided the group with background on UNESCO and on some of the potential opportunities that the organization may provide for the scientific community to consider biosafety and biosecurity. She began by highlighting the *Science Agenda - Framework for Action*[23] that arose from the 1999 World Conference on Science for the Twenty-First Century: A New Commitment, sponsored by UNESCO and the International Council for Science (ICSU). The Conference, held in Budapest, Hungary, provided guidelines for actions addressing: (a) Science for Knowledge; Knowledge for Progress; (b) Science for Peace and Development; and (c) Science in Society and Science for Society. Dr. Hoareau noted UNESCO's role in providing scientific assistance and engaging with policy makers and governments, as well as in working to achieve knowledge transfer and network

[23] UNESCO (United Nations Educational, Scientific, and Cultural Organization) and International Council for Science. 1999. Science Agenda Framework for Action. Adopted by the World Conference on Science, July 1. Available at: *http://www.unesco.org/science/wcs/eng/framework.htm.* Accessed December 11, 2008.

building. She also emphasized the interdisciplinary nature of many of the activities in which UNESCO is engaged and the collaborations that arise between various UNESCO divisions, UN sister agencies such as the FAO, the WHO, and partners in the nongovernmental community.

Dr. Hoareau highlighted several UNESCO programs that might provide opportunities to consider topics in biosecurity, bioethics, and biosafety. For example, the UNESCO Division of Science Policy and Sustainable Development works on policy guidelines and methodologies for the formulation of science policy, particularly to support sustainable development and peace. The Division of Basic and Engineering Sciences also maintains the International Basic Sciences Program, a platform for international cooperation; its aim is to strengthen national capacities in the basic sciences and science education. The ethics of science and technology is also a priority theme for UNESCO. John Crowley of UNESCO's Social and Human Sciences Sector noted that UNESCO incorporates several potentially relevant initiatives including the Bioethics Program, COMEST, and the Global Ethics Observatory databases, which can serve as resources to the community and to member states. Recently, an interagency task group has also been established among WHO, FAO, and UNESCO on biotechnology, and this might provide yet another forum to raise and discuss issues related to biosafety and biosecurity. During the group discussions, it was also pointed out that it could be useful if a statement from scientific bodies was made to the UNESCO Director General that further activities by UNESCO in the area of biosecurity would be relevant.

WHO. Ottorino Cosivi had spoken during the first plenary session about the spectrum of risks posed to global health security in the 21st century and on efforts that WHO has made to support the elimination of chemical and biological weapons and to promote global health security. Although not making a second formal presentation to the working group, Dr. Cosivi further highlighted the need to speak about a range of biological risks and the likelihood that the prioritization of biological risks will vary from country to country. From this starting point more focused efforts could then be made on managing these risks.

OECD. Alexandre Bartsev (OECD) also spoke to the Forum in Plenary 1 on the roles that the OECD has assumed in addressing both biosecurity and emerging technology. Within the working group, Dr. Bartsev contributed several further comments on the ways in which biosecurity may help to create an environment of trust. This could, in turn, help promote industry investment as part of the cycle from basic science research through innovation. Addressing security issues could thus help to provide part of the enabling environment for research and development.

BWC Implementation Support Unit. As a complement to the plenary presentation delivered by Ambassador Avramchev, Chair of the 2008 BWC intersessional process Piers Millett (UN Office for Disarmament Affairs, BWC Implementation Support Unit) highlighted the desire to make the BWC process more inclusive and to continue to incorporate scientific input. He reported that scientists already had roles as members of the national delegations from many of the larger member states. He noted that progress within the BWC has benefited from scientists participating as experts, and pointed to the ability of scientific side events to be organized in conjunction with Convention meetings. International scientific bodies and NGOs may also attend sections of Convention meetings as observers, although Dr. Millett cautioned that there could be the perception that some NGOs might come with their own agendas, which could make member states suspicious of their motives and could make it harder to achieve goals.

Biotechnology Research Center, Tripoli, Libya. The group heard a brief presentation from Mohamed Sharif (Biotechnology Research Center). Libya has partnered with UNESCO and recently established the Biotechnology Research Center, as well as a Bioethics and Biosafety Committee. The Center has initiated collaborations with laboratories and institutions in other countries and was holding national conferences and training programs, while also focusing on issues of laboratory safety. Dr. Sharif highlighted the growth of the biological sciences around the world, and the value that counties with less-developed biological sciences initiatives derived from international collaborations as they worked to build their programs, and the need to provide training in both biological techniques and in laboratory safety and ethics.

OPCW and IUPAC. Ralf Trapp discussed the structure of the OPCW, which administers the Chemical Weapons Convention, and how science advising works in this context. The OPCW includes a Scientific Advisory Board composed of experts from States Parties to the Convention, and this provides an integrated mechanism to feed scientific input directly into the Convention review conferences. However, there was a desire to extend the source of science advice beyond the Scientific Advisory Board and to incorporate expert perspectives from the broader chemistry community. IUPAC, as a neutral, international, nongovernmental body of chemists, was thus able to effectively partner with the OPCW and has hosted two workshops on trends in science and technology relevant to the CWC. One workshop was held in 2002 prior to the First Review Conference, and one was held in 2007 prior to the Second Review Conference (see Chapter 1). Dr. Trapp's remarks pointed out an opportunity in which an international science union was able to provide advice to a policy community in the context of a treaty organization. The partnership also catalyzed an internal process within IUPAC that has led to many further activities addressing issues related to dual use of chemicals and scientific responsibility.

Uganda National Academy of Science (UNAS). Patrick Rubaihayo (UNAS) spoke to the group about a regional workshop UNAS organized in March 2008 on biosafety and security in the life sciences and on providing the opportunity for African scientists to have a voice in such discussions. In surveying existing Ugandan laws on biosafety, UNAS found that they did not address biosecurity concerns. The workshop raised the question of whether countries in East Africa need to adapt existing safety laws and/or create new legal and policy frameworks to capture aspects of biosecurity. Issues of enhancing compliance with existing regulations and incorporating education on biosecurity were also raised. The workshop highlighted the need to reach a common understanding of the scope of biosafety and biosecurity. Although laboratory biosafety and biosecurity are required, workshop participants felt that the primary security risk within Africa arises from natural sources such as disease outbreaks, rather than from research facilities. The issue of intellectual property rights and concerns of biopiracy also loomed large for many African scientists, because of the lack of capacity on the continent and the need to form partnerships with more developed countries.

Dr. Rubaihayo explained that the African participants wished to implement safety and security curricula and standards quickly in order to catch up with the developed world, but lacked infrastructural, human, and financial capacity; and they would need assistance in achieving these goals. He felt that it might be particularly valuable for the developed world to create educational and training materials that could be shared with the developing world to facilitate this process, and that the African science academies should assume more prominent roles in spearheading safety and security awareness and in advising their governments.

Royal Netherlands Academy of Arts and Sciences (KNAW). Finally, Koos van der Bruggen (KNAW) spoke to the working group participants on biosecurity activities that had been undertaken by KNAW. The KNAW has served as the lead academy for the IAP Biosecurity Working Group and played an active role in the formulation of the 2005 IAP Statement on Biosecurity. Following the release and dissemination of this statement, the Dutch Ministry of Science asked KNAW to prepare a code of conduct on biosecurity for scientists and organizations involved with dual use research in the Netherlands. In preparing this code, KNAW held extensive discussions with stakeholders and produced a document of principles that could be translated by each particular organization into its own appropriate context. The code was published in October 2007 and is available online.[24] Follow-on activities such as presentations, articles, and a movie are being prepared. Dr. van der Bruggen explained that, although such a code did not replace existing laws and might not prevent intentionally malicious behavior, it can serve as a useful tool to raise awareness and stimulate discussions. Participants in working group 3 commented that a theme that had emerged from several of the group presentations was that the process of developing a product related to ethical principles could sometimes be even more valuable than the content of the final product.

Working Group Discussions

The working group discussed questions and issues raised by the presentations, as well as reflected on several suggested questions that were posed to the group to help initiate dialogue on these topics. These questions were:

1. What are the different ways in which scientific groups can provide scientific advice on issues related to biosecurity? Which organizations might be interested in having input from the scientific community and where are there such opportunities?
2. What are some examples in which the scientific community has been able to provide advice on biosecurity-related topics to other governmental and nongovernmental, national, and international groups? How did these opportunities arise and how can they be built upon? What were the challenges and lessons learned?
3. Where are there unmet needs and are there ways that the scientific community could start moving to help address these?

Starting Points on the Role of Science Advising. The working group took as its common starting point that the scientific community should provide advice about how to deal with the benefits and potential risks of advances in biology, biotechnology, and the life sciences, including biosecurity matters. Such advice should begin by highlighting the benefits of scientific development and should also be provided within a wider context of biosafety for the following purposes:

---

[24] Royal Netherlands Academy of Arts and Sciences. 2007. A Code of Conduct for Biosecurity. Report by the Biosecurity Working Group. Amsterdam: Royal Netherlands Academy of Arts and Sciences. The code is available at: *http://www.knaw.nl/cfdata/publicaties/detail.cfm?boeken_ordernr=20071092.*

- To build consensus on key issues within the science community, promote proper scientific/ethical conduct, and prevent hindrances of scientific progress;
- To advise policy makers (in different policy areas) on benefits and potential risks, and in this context, on sensible and necessary courses of action;
- To inform, educate, and engage with the public about the risks, and about what is, or should be, done to manage these risks.

The group also considered some of the general aspects that will be needed to provide effective science advice. The participants agreed that science advice happens at different levels, from the personal to the institutional to the national level, as well as regionally and internationally. Thus, the messages that come from the scientific community need to be sincere, consistent, evidence-based, and targeted to the intended audience(s). Effective and relevant policy advice from the science community presents concrete national, regional, and international political strategies and objectives. Science advice will need to be tailored to the expectations, perceptions, experiences, needs, priorities, and political desires of a given context. The science community will also need to get the additional segments of the community, including politicians, parliamentarians, and the general public on board.

Context for Science Advice on Biosecurity. Group discussions returned several times to the varying definitions and interpretations of the term biosecurity. However, the group agreed that biosecurity can be broadly understood as an integrated and interdisciplinary approach to manage biological risks. Biosecurity is therefore about risk perceptions, risk assessment, risk management and risk communication. Science advice has a role to play in all of these areas and scientists need to be involved as part of the policy-shaping processes.

For science advice to be effective, however, it is often necessary to be clear about what is meant by "biosecurity," since the term means different things to different communities. Within the context of the working group discussions at the Forum, the group agreed that the term was referring to a particular set of measures to address the risks emanating from the life sciences, and in particular was addressing scenarios where large numbers of people, animals, plants, or significant parts of the environment, are at risk. It was also understood that the concept of biosecurity is not limited to issues relating to biological weapons or bioterrorism, but must proceed from the recognition of the existing biological risks under given circumstances.

It was felt that the argument for enhancing biosecurity needed to build on well-known historical examples of the risk of abuse of the life sciences for malign purposes, which might include the history of biological weapons and of past biological weapons programs. However, science advice must also account for other evidence (e.g., the cross-border spread of particular animal or plant diseases with severe economic impact), and both current national and regional perceptions and strategies. Within Africa, for example, biotechnology is seen as a strategic opportunity to address key development challenges such as poverty, population growth, and malnutrition. Science advice on biosecurity should be "packaged" into this context, in order to be taken seriously by policy makers and populations. It was pointed out that biosecurity could be a facilitating condition for innovation cycles and thus for economic development, and, therefore, it should not be viewed solely in terms of cost. An opportunity exists to gain much-needed political support, if biosecurity can be integrated into the wider policies of developing countries toward achieving the Millennium Development Goals.

In 2007, for example, the Ministers of Foreign Affairs of Brazil, France, Indonesia, Norway, Senegal, South Africa, and Thailand issued the Oslo Ministerial Declaration—Global Health: A Pressing Foreign Policy Issue of Our Time, as part of their initiative on Global Health and Foreign Policy.[25] The declaration recognized the importance of health issues in policy discussions and the interplay of health with other challenges; the theme on "Capacity for Global Health Security" included item 7.2, namely: "Recognize that the potential of biotechnologies to help developing countries achieve the Millennium Development Goals should not be eclipsed by otherwise legitimate security concerns: establish robust governance mechanisms to prevent misuse of the biological sciences, without hindering their positive contribution to development."

The working group emphasized that advice on biosecurity needed to be multidisciplinary and multisectoral, and had to appreciate that biosecurity is a multistakeholder issue, and hence has to be inclusive. At the international level, this requires coordination and collaboration among the different organizations that have relevant mandates. This could include the UN system and its specialized agencies, as well as organizations outside the UN family such as the OECD, ICRC or OPCW. At the national level, the involvement of many stakeholders in government, science, industry, the NGO community and civil society at large is required, and the communication barriers between these different actors have to be broken down. The group also agreed that, since the responsibilities for

[25] Ministers of Foreign Affairs of Brazil, France, Indonesia, Norway, Senegal, South Africa, and Thailand. 2007. Oslo Ministerial Declaration—Global Health: A Pressing Foreign Policy Issue of Our Time. *Lancet* 369(9570):1373-1378.

biosecurity exist at the levels of the individual, the institution, and nationwide, advice should be targeted to the respective audience(s). Measures to deal with the risks should therefore be complementary, should address ethical matters as well as proper professional conduct more generally, and should be complemented by regulatory instruments and guidelines.

Ways and Means of Science Advising. The group noted that a variety of ways of providing effective scientific advice have been developed and can operate at the several different levels required.

At the national level, advice is being provided by science academies, professional scientific associations and societies, expert committees, and national commissions and advisory boards (e.g., national science and technology ethics commissions, research policy committees). In addition, the working group suggested that scientists should be directly included in national delegations attending negotiations in the area of biosecurity, or areas that are relevant to it. At the national level, it is important for the scientific community to be involved in the review of existing regulatory frameworks within which biosecurity objectives can be accomplished, and for the scientific community to participate in any necessary adaptation of existing regulations and guidelines, or in the creation of additional regulatory mechanisms. It was felt by the group that such reviews will need to be repeated and updated periodically to take account of new developments.

At the regional level, science advice is needed when regional priorities, policies, and capacity-building projects are being discussed and implemented. Regional organizations are important in shaping effective policies and in organizing regional collaborations. Biosecurity should be incorporated into the policy agendas of regional organizations, and regional resources and capacities in the field of biosecurity should be enhanced.

At the international level, a number of organizations have mandates with regard to providing, or facilitating the provision of, advice on biosecurity, and these groups may also facilitate capacity building. These include UNESCO, WHO, FAO, OIE, the International Cooperative Biodiversity Groups, the United Nations Environment Program, the BWC Implementation Support Unit, OECD, and others. International organizations, including specialized agencies, can play important roles with regard to involving the scientific community and in seeking their advice, and providing the governments of their member states with advice based on sound scientific principles and evidence.

At the academic level, organizations such as ICSU, the IAP, and the Academy of Sciences for the Developing World (TWAS) bear a specific responsibility for developing and channeling science advice. International disciplinary science unions also have important roles to play, given their wide geographical participation and legitimacy. Science unions, as well as international scientific bodies such as ICSU, IAP, and TWAS, can help create broad consensus within the scientific community itself. This is essential for consistent and relevant advice to policy makers, as well as for outreach and education directed at the scientific community and the public. The working group noted that international scientific consensus does not necessarily exist on the advantages and risks created by developments in the life sciences. Such dialogues among the scientific community should be played out not only within the policy sphere, but also addressed within the international scientific community as it moves toward achieving a level of consensus.

Unions as well as ICSU, IAP, and TWAS can also help promote common standards (including on professional ethics), foster the education of future generations of scientists and engineers, and inform both policy makers and the public. They can do this in collaboration with other international agencies, but equally important is their ability to work through their own national constituent bodies to transmit these messages in a tailored and relevant fashion. Unions, as well as interacademy bodies, also can be effective channels to involve industry in the development of policy advice.

Treaty-based institutional mechanisms such as the Scientific Advisory Board of the OPCW, or national governmental science advisory bodies involved in the CWC context, or in the BWC processes, have also been effective. The involvement of scientists as delegation members, or by serving in capacities such as members of NGOs, scientific associations, or as individuals, has proven useful.

The group noted that there is a need for effective and targeted outreach and communication of biosecurity issues. Given the diverse audiences seeking or requiring advice, the variety of publications and other communications (e.g., press, electronic media, and the Internet) ought to be tailored to these different audiences. Such audiences will include scientists, media, policy makers, and the "general public." There is also a need for education and training programs, including education and training for practicing scientists and other practitioners, training at the university level for upcoming generations of scientists, and also education for policy and law-makers. It would be desirable to share existing resources such as training materials, educational videos, and other tools on a wider basis.

The working group also recognized that resources may need to be devoted to assist developing countries in building their capacities to provide scientific advice.

Some Pitfalls. The working group noted that, at the moment, there appears to be a lack of coordination among the various efforts to address biosecurity. There is no coherent international strategy, and a lack of collaboration among the different actors. There is a danger that efforts are being duplicated and a stock-taking exercise would be desirable to review which efforts are actually under way, and how effective they are.

The group also emphasized that no single international organization can cover all issues related to biosecurity, let alone the overall issues related to risk assessment, management, and communication related to advances in the biosciences. The same applies at the national level. There is a need for coordination, networking, and information sharing. In some instances, for example, interministerial mechanisms may be needed. During the discussions, it was noted that South Africa, for example, is already working to develop networks among groups such as university research directors, and has created the National Science and Technology Forum as a mechanism to bring together some of the relevant constituents.

On the other hand, it must be understood that enhancing and enforcing regulatory frameworks, providing science advice, adopting ethical codes, and providing education and outreach can achieve only so much, and that these efforts cannot and should not be expected to completely deter or prevent acts of malevolence.

Results

The working group proposed the following four suggestions.

1. There is a need for better coordination at the international level. The United Nations should facilitate this and take the lead; it can and should bring together the major stakeholders, including industry, the scientific community, civil society and governments, into a common program aimed at ensuring that advances in the life sciences are used only for the benefit of humankind. Under a broad umbrella such as the UN could provide, it would be easier to synchronize the diverse and multifaceted efforts of specialized agencies, organizations such as the OECD, and many other international actors and to address these issues on the basis of well-established interagency coordination mechanisms.

2. Consideration should be given to the organization of sessions, side-events or other forums on biosafety and biosecurity issues in the context of forthcoming meetings. Some examples of possible opportunities include:

- World Conference on Science (Budapest + 10) in late 2009;
- Global Ministerial Forum on Research for Health (Bamako, Mali) in November 2008;
- UNESCO World Commission on the Ethics of Scientific Knowledge and Technology (extraordinary session November 2008 in Paris, France, and an ordinary session June 2009 in Singapore);
- UNESCO International Bioethics Committee (ordinary session November 2008 in Paris, France);
- World Social Science Forum of the International Social Science Council (Bergen, Norway) in May 2009.

3. There is a need for improved networking and for building networks of networks at national, regional and international levels. Different communities that have a contribution to make to biosecurity should be brought together, including life sciences, security and law enforcement, policy makers, lawyers, and others. ICSU, IAP, and TWAS should take the lead to create such networks of networks. Science unions should get involved as well, and can work through their national constituencies to promote biosecurity in the local/regional context and within a broader perspective on risk assessment, management, and communication regarding advances in the life sciences.

4. The existing connecting points between science and policy making at the national levels should be used and, where necessary, should be energized, in order to promote better communication and cooperation between the scientific and policy communities.

Highlights of the group discussions and these four results were presented by Ralf Trapp to the entire Forum in a plenary session.

3

Major Themes and Next Steps

CONCLUDING PLENARY DISCUSSIONS

Following the presentation and discussion of the working group reports (Plenary 6), the concluding plenary session (Session 7) of the meeting allowed Forum participants to discuss overall themes that had arisen and to look forward to potential next steps. Michael Clegg, Foreign Secretary of the U.S. National Academy of Sciences and chair of the Forum Oversight Committee, opened this final session with a reflection on some of the goals he hoped that the International Forum on Biosecurity would achieve. These included:

- To assess progress since the 1st International Forum (held in 2005);
- To discuss the roles and responsibilities of the international scientific community in fostering policies that promote both continuing scientific progress and greater international security;
- To foster communication and cooperation among the leaders from the international scientific community and among other stakeholders in the biosecurity field;
- To consider challenges to biosecurity from emerging disciplines and their regulation; and
- To provide input from the international scientific community into the preparations for the 2008 Biological and Toxin Weapons Convention (BWC) Meeting of Experts and other biosecurity initiatives.

Dr. Clegg also spoke of several roles that the international scientific community might play, including educating scientists to foster a sense of a culture of responsibility, protecting the practice and progress of legitimate science, and providing advice to governments on technical issues. Discussion then followed among all of the Forum participants, as they reflected on the two-and-a-half day meeting.

MAJOR THEMES

Several themes and concerns emerged repeatedly across the working group and plenary discussions.

Participants in the Forum agreed that progress in the life sciences is global in nature and thus that a dialogue on issues such as the risks that may arise from potential misuse of life sciences, and developments and strategies to minimize such risks, must also be international. However, there are challenges associated with the use of the terms "biosecurity" and "dual use," which mean many different things to different groups. Many Forum participants emphasized that it is more effective to present ideas in more easily understood terms, and that it is important to be as specific as possible about what is meant in each particular context.

Many of the presentations and discussions during the Forum also highlighted the concept that biosecurity and dual use issues in the life sciences were fundamentally about *risk*, specifically about assessing relative risks and developing appropriate risk management options. Furthermore, many participants agreed that achieving a balance among potential risks is critical. They felt that it is important to acknowledge and to take steps to minimize risks of the life sciences being misused for harm, to recognize the serious risks to human, veterinary, and environmental health that could result from such misuse, and also to minimize the risk of an overemphasis on security, which could stifle progress in the life sciences and therefore could reduce the many important benefits of such progress.

The three working groups also highlighted the need for a global dialogue on these issues, recognizing that there will likely be differences in the priorities that different regions, nations, and segments of the scientific and policy communities may place on achieving various objectives. Therefore, developing a toolkit of multiple options for addressing topics such as education, oversight, and governance was widely recognized to be valuable, since different groups may choose to emphasize different strategies.

There was also recognition by Forum participants that consensus does not yet exist within the scientific community on many of the questions considered by the Forum. As the working group on science advising noted, discussions will need to continue within the international scientific community, so that it continues to work toward and build upon areas of consensus. The group felt that providing clear and consistent advice on biosecurity issues to additional stakeholders, such as government policy makers and the general public, would be greatly strengthened if the life sciences community could speak in greater harmony.

The Forum discussions included several examples of areas in which consensus within the international life sciences community does not yet exist, or in which there are distinct differences in priorities. Participants from some regions, particularly in the developing world, expressed the view that naturally occurring pathogens in the environment were a significantly greater subject of biosecurity concern than were threats that might be generated in laboratories. An important issue remains regarding how to incorporate this concern into biosecurity discussions. Many participants highlighted the need to recognize these differing priorities among different groups, and further suggested that packaging discussions of the range of biosecurity issues into whichever contexts would be most relevant and appealing for different groups, would be valuable. On the other hand, other participants reflected on the complications that arise when trying to address multiple aspects of an issue, pointing to the additional international organizations, expertise, and stakeholder groups that would be needed to discuss natural disease hazards. These participants were resistant to broadening the discussions beyond biosecurity as it relates to misuse of research and related activities.

Another subject about which some participants expressed differences of opinion was where the most appropriate balance of oversight lies on the spectrum from self-governance to more formalized regulation. The participants in working group 2, which considered models of oversight, agreed that a balance will be needed between "bottom-up" and "top-down" oversight measures, and this view was widely shared by the Forum. Many Forum participants also recognized that developing an array of possible options might be desirable, and that different groups might choose to select different combinations of options.

Education was a common strategy emphasized by the three working groups to help move toward greater awareness of dual use issues, and ultimately toward greater consensus about risks and risk management strategies within the scientific community. The Forum discussions included suggestions to begin educational efforts by highlighting the many benefits arising from scientific developments, to incorporate specific historical examples of previous misuse of science, and also to promote active thinking and learning about biosecurity. A number of participants suggested that States Parties to the BWC should commit to taking steps to advance education and that national and international scientific organizations should promote the need for biosecurity education as well. The engagement of multiple stakeholders in the creation of codes of conduct was seen by many workshop participants as one opportunity to further such educational objectives. Beyond the creation of codes of conduct, participants suggested that discussions of the potential risks of misuse from life sciences advances, responsible conduct of science, and the existence of the BWC should be incorporated into academic training programs, although there was recognition that this would be a difficult task.

The need to communicate and collaborate was another important theme arising from the workshop discussions. It was understood by the participants that building networks, sharing information about efforts and initiatives already under way or contemplated, and sharing resources that had been developed, is of great value to the community. Discussions reflected the opinion that no single organization or group can address all aspects related to biosecurity and biosafety, and that participants could all benefit from the results and materials produced by the various groups and initiatives tackling aspects of these topics. Similarly, many workshop participants highlighted the need for evaluation of the efforts that were going forward, and particularly for more information on what had been most successful in different contexts. Participants from less developed countries in particular expressed a desire for access to shared products and materials that might help them rapidly advance the development of biosafety and biosecurity standards in more resource-limited regions. One idea that emerged from the discussion in working group 1 was to create a clearinghouse of materials and information for wide use by the community and this idea received substantial support during the plenary discussions.

Throughout the Forum, many of the individual participants as well as the three working groups, highlighted the critical issue of funding. Financial support is needed to convene meetings to discuss biosecuity topics, to develop and deliver educational programs such as seminars, and to maintain networks and forums for sharing information. It is not clear where such funding will come from, and this is likely to remain a concern.

GREATER COLLABORATION AMONG INTERNATIONAL SCIENTIFIC ORGANIZATIONS

In parallel with the emphasis on continuing the dialogues within the scientific community on biosecurity and biosafety, was an emphasis by many Forum participants on the convening powers of various international scientific organizations. Within the nongovernmental community, the Forum participants pointed to organizations such as the International Council for Science (ICSU), the InterAcademy Panel on International

Issues (IAP), the Academy of Sciences for the Developing World (TWAS), and the disciplinary science unions. These groups are often seen to have legitimacy as neutral networks that can engage scientists from many countries. In addition, they can take advantage of their networks of national member bodies to help exchange opinions and disseminate information.

The suggestion was raised by many participants at the Forum that it would be particularly useful if IAP and ICSU could strengthen their collaboration to continue dialogues on biosecurity issues within the scientific community. The IAP currently has a Biosecurity Working Group, and working group 1 suggested that IAP convene a task group to consider the specifics of how best to establish a clearinghouse for resources such as educational materials, and to plan additional follow-on initiatives. The IAP Biosecurity Working Group could thus form a valuable locus within IAP to interact with its partners in the scientific community, including with ICSU and the life sciences unions.

On the intergovernmental level, many Forum participants felt that organizations such as the United Nations and the World Health Organization have critical roles to play, and are also in a position to help effectively bridge between communities of scientific experts and governmental agencies. The BWC provides the fundamental norm against the misuse of the life sciences, and the intersessional process has proved to be a valuable convening mechanism to address topics relevant to the scientific community. It was suggested by many participants that the UN serve as the locus for a broad international forum on biosecurity because of its ability to incorporate multiple stakeholders including the laboratory, human and veterinary health, government, and security communities. The former UN Secretary General, Kofi Annan, had proposed convening such a forum and several Forum participants expressed the hope that this proposal might ultimately move forward.

CONCLUSION AND ADJOURNMENT OF THE FORUM

Dr. Clegg closed the Forum by highlighting the value of the process of discussion by the international life sciences community on topics of biosecurity and biosafety. He thanked all of the participants for their active engagement, and looked forward to continuing efforts by the community to address these issues.

References

Annan, K. 2006. Uniting Against Terrorism: Recommendations for a Global Counter-Terrorism Strategy. Report of the Secretary-General. A/60/825. New York: United Nations. Available at: *http://www.un.org/unitingagainstterrorism/contents.htm.*

Atlas, R., and M. Dando. 2006. The dual use dilemma for the life sciences: Perspectives, conundrums, and global solutions. *Biosecurity and Bioterrorism: Biodefense Strategy, Practice, and Science* 4(3):276-286.

Balali-Mood, M., P.S. Steyn, L.K. Sydnes, and R. Trapp. 2008. Impact of scientific developments on the Chemical Weapons Convention (IUPAC Technical Report). *Pure and Applied Chemistry* 80(1):175-200. Available at: *http://www.iupac.org/publications/pac/80/1/0175/.*

Biological Weapons Convention Meetings Secretariat. 2006. Sixth Review Conference of the States Parties to the Biological Weapons Convention. Final Document. Geneva: United Nations Department for Disarmament Affairs. Available at: *http://www.opbw.org.*

Bügl, H., J.P. Danner, R.J. Molinari, J. Mulligan, D.A. Roth, R. Wagner, B. Budowle, R.M. Scripp, J.A.L. Smith, S.J. Steele, G. Church, and D. Endy. 2006. A Practical Perspective on DNA Synthesis and Biological Security. International Consortium for Polynucleotide Synthesis. December 4. Available at: *http://pgen.us/ICPS.htm.*

Carter, A., J. Deutch, and P. Zelikow. 1998. Catastrophic terrorism: Tackling the new danger. *Foreign Affairs* 77(6):80-94.

Epstein, G.L. 2001. Controlling biological warfare threats: Resolving potential tensions among the research community, industry, and the national security community. *Critical Reviews in Microbiology* 27:321-354.

Fischer, J.E. 2006. Stewardship or Censorship: Balancing Biosecurity, the Public's Health, and the Benefits of Scientific Openness. Washington, DC: Stimson Center. Available at: *http://www.stimson.org/globalhealth/pdf/Stewardship.pdf.*

Fox, J.L. 2003. Bioterrorism threat could make some research too "sensitive" to disclose. *ASM News 69(3):112-114.* Available at: *http://www.asm.org/microbe/index.asp?bid=13147.*

Frerichs, R.L., R.M. Salerno, K.M.Vogel, N.B. Barnett, J. Gaudioso, L.T. Hickok, D. Estes, and D.F. Jung. 2004. Historical Precedence and Technical Requirements of Biological Weapons Use: A Threat Assessment. SAND2004-1854. Albuquerque, NM: Sandia National Laboratories.

Garfinkel, M.S., D. Endy, G.L. Epstein, and R.M. Friedman, eds. 2007. Working Papers for Synthetic Genomics: Risks and Benefits for Science and Society. Available at: *http://hdl.handle.net/1721.1/39658.*

Jackson, R.J., A.J. Ramsay, C.D. Christensen, S. Beaton, D.F. Hall, and I.A. Ramshaw. 2001. Expression of mouse interleukin-4 by a recombinant ectromelia virus suppresses cytolytic lymphocyte responses and overcomes genetic resistance to mousepox. *Journal of Virology* 7(3):1205-1210.

Journal Editors and Authors Group. 2003a. Uncensored exchange of scientific results. *Proceedings of the National Academy of Sciences* 100(4):1464.

Journal Editors and Authors Group. 2003b. Statement on the consideration of biodefense and biosecurity. *Nature* 421:771.

Journal Editors and Authors Group. 2003c. Statement on scientific publication and security. *Science* 299(5610):1149.

Khan, M. 2006. Preparations and expectations. Presentation to the United Nations General Assembly First Committee. Sixth Review Conference of the Biological and Toxin Weapons Convention: New York: United Nations, October 11. Available at: *http://www.unog.ch/80256EDD006B8954/(httpAssets)/298DFC7CC2CD636BC125720D0045B3C8/$file/First_Committee_BWC_thematic_presentation_slides.pdf.*

Leitenberg, M. 2005. Assessing the Biological Weapons and Bioterrorism Threat. Carlisle Barracks, PA: Strategic Studies Institute, U.S. Army War College.

Meselson, M. 2000. The problem of biological weapons. Symposium on Biological Weapons and Bioterrorism, National Academy of Sciences, Washington, DC, May 2.

Miller, J., S. Engelberg, and W. Broad. 2001. *Germs: Biological Weapons and America's Secret War*. New York: Simon and Schuster.

Ministers of Foreign Affairs of Brazil, France, Indonesia, Norway, Senegal, South Africa, and Thailand. 2007. Oslo Ministerial Declaration—Global Health: A Pressing Foreign Policy Issue of Our Time. *Lancet* 369(9570):1373-1378.

NRC (National Research Council). 2004a. *Biotechnology Research in an Age of Terrorism*. Washington, DC: The National Academies Press.

NRC (National Research Council). 2004b. *Seeking Security: Pathogens, Open Access, and Genome Databases*. Washington, DC: The National Academies Press.

NRC (National Research Council). 2006. *Globalization, Biotechnology, and the Future of the Life Sciences.* Washington, DC: The National Academies Press.

OECD (Organisation for Economic Co-operation and Development). 2004. Promoting Responsible Stewardship in the Biosciences: Avoiding Potential Abuse of Research and Resources. Chairman's Summary. Paris: OECD. Available at: *http://www.oecd.org/dataoecd/30/56/33855561.pdf.*

OECD (Organisation for Economic Co-operation and Development). 2007. *OECD Best Practice Guidelines on Biosecurity for BRCs (Biological Resource Centers).* Paris: OECD. Available at: *http://www.oecd.org/dataoecd/6/27/38778261.pdf.*

Parshall, G.W., G.S. Pearson, T.D. Inch, and E.D. Becker. 2002. Impact of Scientific Developments on the Chemical Weapons Convention (IUPAC Technical Report). *Pure and Applied Chemistry* 74(12):2323-2352. Available at: *http://www.iupac.org/publications/pac/2002/7412/index.html.*

Pearson, G.S. 1993. Prospects for chemical and biological arms control: The web of deterrence. *The Washington Quarterly* 16(Spring):145-162.

Rappert, B. 2004. Towards a Life Science Code: Countering the Threats from Biological Weapons. *Bradford Briefing Paper* No. 13. Available at: *http://www.brad.ac.uk/acad/sbtwc.*

Reppy, J. 2007. The end of dual use? Implications for export control policy. Paper prepared for presentation at the 48th Annual Meeting of the International Studies Association, Chicago, IL. March. Available at: *http://www.allacademic.com/meta/p_mla_apa_research_citation/1/7/8/8/3/p178830_index.html.*

Ripandelli, D. 2005. Building blocks for a code of conduct for scientists, in relation to the safe and ethical use of biological sciences. Presentation to the 2005 Meeting of Experts of the Biological Weapons Convention. Geneva. June 13. Available at: *http://www.opbw.org/.*

Rossiskiye Vesti. 1992. Interview with President Boris Yeltsin. Washington, DC: Foreign Broadcast Information Service, FBIS-SOV-92-103, May 27.

Royal Netherlands Academy of Arts and Sciences. 2007. A Code of Conduct for Biosecurity. Report by the Biosecurity Working Group. Amsterdam: Royal Netherlands Academy of Arts and Sciences. Available at: *http://www.knaw.nl/cfdata/publicaties/detail.cfm?boeken__ordernr=20071092.*

Royal Society. 2006. Report of the RS-IAP-ICSU International Workshop on Science and Technology Developments Relevant to the Biological and Toxin Weapons Convention. London: Royal Society. Available at: *http://royalsociety.org/displaypagedoc.asp?id=22789.*

Tumpey, T.M., C.F. Basler, P.V. Aguilar, H. Zeng, A. Solórzano, D.E. Swayne, N.J. Cox, J.M. Katz, J.K. Taubenberger, P. Palese, and A. García-Sastre. 2005. Characterization of the reconstructed 1918 Spanish influenza pandemic virus. *Science* 310(5745):77-80.

United Nations. 2006. The United Nations Global Counter-Terrorism Strategy. UNGA Resolution A/RES/60/288. New York: United Nations, Annex, II–11. Available at: *http://www.un.org/terrorism/strategy-counter-terrorism.shtml#resolution.*

UNESCO (United Nations Educational, Scientific, and Cultural Organization) and International Council for Science. 1999. Science Agenda Framework for Action. Adopted by the World Conference on Science, July 1. Available at: *http://www.unesco.org/bpi/science/wcs/eng/framework.htm.*

Vogel, K.M. 2006. Bioweapons proliferation: Where science studies and public policy collide. *Social Studies of Science* 36(5):659-690.

Vogel, K.M. 2008. Framing biosecurity: An alternative to the biotech revolution model? *Science and Public Policy* 35(1):45-54.

WHO (World Health Organization). 2004. *Laboratory Biosafety Manual,* 3rd ed. Geneva: WHO. WHO/CDS/CSR/LYO/2004.11. Available at: *http://www.who.int/csr/resources/publications/biosafety/WHO_CDS_CSR_LYO_2004_11 /en/.*

WHO (World Health Organization). 2005. Life Science Research: Opportunities and Risks for Public Health. Geneva: WHO. WHO/CDS/CSR/LYO/2005.20. Available at: *http://www.who.int/csr/resources/publications/deliberate/WHO_CDS_CSR_LYO_2005_20/en/index.html.*

WHO (World Health Organization). 2007. Scientific Working Group on Life Science Research and Global Health Security: Report of the First Meeting. Geneva: WHO. WHO/CDS/EPR/2007.4. Available at: *http://www.who.int/csr/resources/publications/deliberate/WHO_CDS_EPR_2007_4.*

Wimmer, E. 2006. The test-tube synthesis of a chemical called poliovirus. *EMBO Reports* 7(Special Issue):S3-S9.

Zilinskas, R., and J.B. Tucker. 2002. Limiting the contribution of the open scientific literature to the biological weapons threat. *Online Journal of Homeland Security* (December). Available at: *http://www.homelandsecurity.org/journal/Articles/.tucker.html.*

Appendixes

A

Committee Member Biographies

Michael T. Clegg (NAS), (*Chair*), is the Donald Bren Professor of Biological Sciences at the University of California, Irvine, and foreign secretary of the National Academy of Sciences. He also serves as the chair of the National Academy of Sciences Council Committee on International Affairs. His research specialty is population genetics and molecular evolution, particularly the study of evolutionary genetic changes in plant populations. Prior to moving to Irvine, he was professor of genetics at the University of California, Riverside. He served as dean of the College of Natural and Agricultural Sciences, University of California, Riverside from 1994 to 2000, and he was the founding director of the Genomics Institute, now renamed the UCR Institute for Integrative Genome Biology. He received his B.S. in agricultural genetics and Ph.D. in genetics at the University of California, Davis. Dr. Clegg has received numerous awards including a Guggenheim Fellowship (1981) and the Darwin Prize of Edinburgh University (1995). Dr. Clegg was elected to membership in the U.S. National Academy of Sciences in 1990, and he was elected a fellow of the American Academy of Arts and Sciences in 1992. He served as president of the American Genetic Association (1987), president of the International Society for Molecular Biology & Evolution (2002), and chair of the Section on Agriculture, Food and Natural Resources of the American Association for the Advancement of Science (2003).

Gail H. Cassell (IOM) is Vice President for Scientific Affairs and Distinguished Lilly Research Scholar for Infectious Diseases, Eli Lilly and Company in Indianapolis, Indiana. She is the former Charles H. McCauley Professor and chairman of the Department of Microbiology at the University of Alabama School of Medicine and Dentistry at Birmingham. Dr. Cassell was a member of the National Institutes of Health (NIH) Director's Advisory Committee and a member of the Advisory Council of the National Institute of Allergy and Infectious Diseases of NIH. She was named to the original Board of Scientific Councilors of the Center for Infectious Diseases, Centers for Disease Control and served as chair of the Board. She recently served a three-year term on the Advisory Board of the Director of the Centers for Disease Control and is a member of the Secretary of Health and Human Services Advisory Council of Public Health Preparedness. Currently she is a member of the Science Board of the Federal Food and Drug Administration. Since 1996, she has been a member of the U.S.-Japan Cooperative Medical Science Program responsible for advising the respective governments (U.S. State Department/Japan Ministry of Foreign Affairs) on joint research agendas. She has served on several editorial boards of scientific journals and has authored over 250 articles and book chapters. Dr. Cassell has received national and international awards and an honorary degree for her research in infectious diseases. She is a member of the Institute of Medicine (IOM) of the National Academies and is currently serving on the IOM Council. She is a past president of the American Society for Microbiology (ASM) and has served as chair of the ASM Public and Scientific Affairs Board, is a former vice-president of the International Union of Microbiological Societies, has served as an advisor to the White House Office of Science and Technology Policy, and has been an invited participant in numerous congressional hearings and briefings related to infectious diseases, antimicrobial resistance, and biomedical research. She has served two terms on the Liaison Committee on Medical Education, the accrediting body for U.S. medical schools as well as other national committees involved in establishing policies for training in the biomedical sciences. She is currently a member of the Executive Committee of the Board of Visitors of Columbia University School of Medicine, the Board of Directors of the Burroughs Wellcome Fund, the Advisory Council of the School of Nursing of Johns Hopkins University, and an Emeritus Director of Research!America.

Hernan Chaimovich is professor in the Department of Biochemistry of the Institute of Chemistry, Universidade de São Paulo (USP), Brazil. He received a degree in Biochemistry from the Universidad de Chile, worked with Osvaldo Cori in apyrase enzymology, and spent two and a half years in the United States working under the supervision of C.A. Bunton (University of California, Santa Barbara) and F.H. Westheimer (Harvard University) in physical organic chemistry. He returned to Chile as an assistant professor of Biochemistry and moved to Brazil in 1969, first as a fellow of the Fundação de Amparo à Pesquisa do Estado de São Paulo in the Department of Physiology at the Faculty of Medicine, Universidade de São Paulo (USP) and later in the Department of Biochemistry of the Institute of Chemistry, USP, where he became a full professor in 1985. Dr. Chaimovich's research is in interfacial effects on chemical and biological reactivity using micelles and vesicles as models. The contributions of his group, including theoretical and experimental studies of the effect of micelles and vesicles on a number of chemical reactions, have contributed to dissecting the effect of these aggregates on chemical reactivity. Dr. Chaimovich participated in the establishment of an interdisciplinary graduate course in Biotechnology and an undergraduate Course in Molecular Sciences, which he coordinated for four years. He was a director of the Teachers Association of USP. An active member of the Brazilian Society of Biochemistry and Molecular Biology, he was elected president of the Society in 1994. In recognition of his academic contributions, he was elected a fellow of the American Association for the Advancement of Science and a member of the Ordem Nacional do Mérito Científico e Tecnológico. Dr. Chaimovich has recently served as vice-president for external relations for the International Council for Science (ICSU).

Roderick Flower, FRS, is professor of Biochemical Pharmacology at the William Harvey Research Institute, Queen Mary, University of London. Dr. Flower received his postgraduate training at the department of pharmacology in the Royal College of Surgeons of England under the supervision of Sir John Vane and subsequently worked as part of the prostaglandin research team at the Wellcome Foundation in Kent until 1984. Dr. Flower served as Chair of Pharmacology at the University of Bath, where he also served as head of the School of Pharmacy and Pharmacology (1987-1989). In 1989, he moved to the medical college of St. Bartholomew's Hospital, where he became a director and founding member of the William Harvey Research Institute, and started a new Department of Biochemical Pharmacology. He served as head of the Institute (1998-2002) and was also Wellcome Principal Research fellow (1994-2007). His main interests are the mechanism of action of anti-inflammatory drugs including Cox inhibitors and glucocorticoid steroids. Dr. Flower has published over 200 peer-reviewed research papers and holds several patents. He has trained numerous Ph.D. students, hosted many researchers from overseas in his group and made important contributions to undergraduate teaching. He has served on editorial and scientific boards and was president of the British Pharmacological Society (2000-2003). Dr. Flower's honors include the Thomas Woodcock Physiology Prize (University of Sheffield 1972), the British Pharmacological Society's Sandoz Prize (1978), the Gaddum Memorial Lecture and Medal of The British Pharmacological Society (1986), Fellow of the Academy of Medical Sciences (1999), Fellow of Academia Europeae (2001), the William Withering Medal of the Royal College of Physicians (2003), the Lifetime Achievement Award of the International Society of Inflammation Associations (2005), and the Bayliss-Starling Prize Lecture of the Physiological Society (2006). He was elected a fellow of the Royal Society in 2003 and served on the international organizing committee for a workshop hosted by the Royal Society, the InterAcademy Panel on International Issues, and ICSU in 2006 on implications of life sciences developments for the Biological and Toxin Weapons Convention.

David Franz is chief biological scientist at the Midwest Research Institute. He served in the U.S. Army Medical Research and Materiel Command for 23 of his 27 years on active duty. Dr. Franz has served as both deputy commander and then commander of the U.S. Army Medical Research Institute of Infectious Diseases and as deputy commander of the U.S. Army Medical Research and Materiel Command. Prior to joining the Command, he served as group veterinarian for the 10th Special Forces Group (Airborne). Dr. Franz served as chief inspector on three United Nations Special Commission biological warfare inspection missions to Iraq, and as technical advisor on long-term monitoring. He also served as a member of the first two United States/United Kingdom teams that visited Russia in support of the Trilateral Joint Statement on Biological Weapons, and as a member of the Trilateral Experts Committee for biological weapons negotiations. Dr. Franz has served on numerous National Research Council (NRC) committees working on biosecurity-related issues in the United States and internationally, and was a member of the committee that produced the NRC report *Biotechnology Research in an Age of Terrorism* (2004). He is currently a member of the National Science Advisory Board for Biosecurity and is co-chair of its international collaborations working group.

Andrzej Górski is professor of medicine and immunology at the Medical University of Warsaw and vice president of the Polish Academy of Sciences. He is board certified in internal medicine with a subspecialty certification in clinical immunology. Dr. Górski received his M.D. (1970) and Ph.D. (1973) degrees from the Medical University of Warsaw and was a Fulbright Scholar at the Sloan-Kettering Institute for Cancer Research. He has been a visiting professor at Adelaide Children's Hospital, Australia, the Weizmann Institute of Science, Israel, the University of London United Medical and Dental Schools of Guy's and St. Thomas's Hospitals, England, and the Universidad Autonoma, Madrid, Spain. Dr. Górski served as prorector for Scientific Affairs & International Cooperation (1993-1996) and as rector (1996-1999) of The Medical University of Warsaw. From

1999-2007 he was also director of the L. Hirszfeld Institute of Immunology and Experimental Therapy at the Polish Academy of Sciences. Dr. Górski has authored over 100 scientific publications, serves as the editor in chief of *Archivum Immunologiae et Therapiae Experimentalis,* and has served as a member of the editorial board of *Science & Engineering Ethics.* His awards include the Meller Award for excellence in cancer research from the Sloan-Kettering Institute, the ICRETT award and the Yamagiwa-Yoshida award from the International Union Against Cancer, the J. Sniadecki Memorial award from the Polish Academy of Sciences (the highest award in medical sciences in Poland), and the Gloria Medicinae, awarded by The Polish Medical Association. In addition, Dr. Górski is a member of the Committee for Ethics in Science at the Polish Academy of Sciences, a member of the Committee for Ethics in Science at the Ministry of Science, head of the Bioethics Committee, Ministry of Health, and represents Poland in the Forum of National Ethics Committees to the European Commission. He has served as chair of the scientific committee for the workshop "The Advancement of Science and the Dilemma of Dual Use," held under the auspices of the Polish Academy of Sciences and UNESCO in November 2007.

Indira Nath is Director of the Blue Peter Research Centre, Hyderabad, India. The center is part of the LEPRA Society, a leading Indian non-governmental organization working in the fields of leprosy, tuberculosis, malaria, HIV/AIDS, and disabilities. Dr. Nath previously served as dean of the School of Medicine at the Asian Institute of Medicine, Science and Technology in Malaysia. She was formerly S.N. Bose Research Professor, one of the five named national professorships endowed by the Indian National Science Academy in recognition of outstanding achievements in research. Dr. Nath also served as the founder head of the Department of Biotechnology at the All India Institute of Medical Sciences. After receiving her M.D., Dr. Nath worked as a house officer in hospitals in the United Kingdom and participated in a Nuffield Fellowship at London's National Institute of Medical Research. Her area of specialization is pathology with a special interest in the immunology of infectious diseases. Dr. Nath has received numerous awards including Padma Shri (1999), bestowed by the president of India for Individual Contribution in Immunology, Chevalier le National Order of Merite (2003), Docteur Honoris Causa (2003) awarded by the University Pierre and Marie Curie, L'Oreal-UNESCO Women in Science (2002), the Shri Om Prakash Bhasin Foundation Award (1990), and the S.S. Bhatnagar Award of the Council for Scientific and Industrial Research (1983). She has served on several committees of the International Council for Science, including the Committee on Science and Social Responsibility (2003) and the Scoping Group on Health (2006-2007). She has also served as a Country Representative to the UN Commission on Science and Technology for Development (2003-2005), was a member of the Scientific Advisory Committee to the Cabinet (2003), a member of the Steering Committee in Science and Technology for 11th Planning Commission, Government of India (2006), and a member of the Working Group for Formulation of 11th Five Year Plan on Cross Disciplinary Technology Areas (2006). Dr. Nath is a fellow of the Indian National Science Academy, the Indian Academy of Sciences, the National Academy of Sciences (India), the National Academy of Medical Sciences (India), the Academy of Sciences for the Developing World (TWAS), and the Royal College of Pathologists, United Kingdom.

Barbara Schaal (NAS) is the Spencer T. Olin Professor in Arts and Sciences in the Department of Biology at Washington University in St. Louis and vice president of the National Academy of Sciences. She is also chair of the NAS Council Committee on Scientific Programs and chair of the National Academies Division Committee for the Division of Earth and Life Studies. Dr. Schaal was elected to the National Academy of Sciences in 1999 for her investigations into the evolution of plant populations. Her work on the application of DNA analysis to plant evolution at the population level showed an unexpectedly high level of diversity due to limited gene migration. Her research includes the use of gene genealogies and coalescence theory to detect the geographical patterns of gene migration between populations of North American native plants. She also conducts studies on the species relationships in plants native to South America, Africa, and Asia and on issues related to the conservation of rare plants. Her current work examines gene flow and genetic diversity in wild and cultivated Asian rice. Dr. Schaal served as chairperson of Washington University's Department of Biology (1993-1997), chairperson of the Scientific Advisory Council for the Center for Plant Conservation, president for the Society for the Study of Evolution, associate editor for the journal *Molecular Biology and Evolution*, and president of the Botanical Society of America. She received her Ph.D. in Population Biology from Yale University in 1974.

Leiv Sydnes is professor of chemistry at the University of Bergen, Norway, a position he has held since 1993. His research is concentrated on organic synthesis with an emphasis on the application of cyclopropane chemistry and photochemistry to introduce useful structures into organic molecules. He received his Ph.D. in organic chemistry from the University of Oslo in 1978, conducted a postdoctoral fellowship at the University of Western Ontario, Canada, and previously served as an associate professor and professor at the University of Tromsø (1980-1993). He has also been a visiting professor at Iowa State University and a visiting fellow at the Australian National University, Canberra. Dr. Sydnes is past president of the International Union of Pure and Applied Chemistry (IUPAC) and chair of the IUPAC CHEMRAWN (CHEMical Research Applied to Word Needs) Committee. He has served on the organizing committee for two international workshops convened by IUPAC and the Organisation for the Prohibition of Chemical Weapons that have addressed trends in chemical sciences and technology ahead of the First and Second Chemical Weapons Convention Review Conferences. He formerly served on the European Communities Chemistry Council, the European Communities Registration Board, and the Federation of European Chemical Societies, as well as serving as President of the Norwegian Chemical Society (1992-1996). In addition to numerous articles in peer-reviewed scientific journals, Dr. Sydnes has written more than 60 articles in national periodicals, including chemistry topics for nonspecialists and discussion papers on teaching. He has authored 5 books covering laboratory work in introductory and intermediate organic chemistry at the university level, 19 high school chemistry textbooks, and 2 textbooks for chemical colleges, and has contributed chapters to numerous technical or more general scientific reports. Dr. Sydnes was elected to the Norwegian Academy of Science and Letters (1999) and his honors include Kyoto Institute of Technology Lectureship (1990), the Thaulow Prize (1995), and the Federation of the European Chemical Societies Medal (2003). He is also an elected fellow of the Royal Society of Chemistry (2004), an elected member of Academia Europaea (2005), and an elected fellow of the Federation of Asian Chemical Societies (2005).

B

Agenda

Sunday, March 30

All day	Participants arrive
5:30PM	Registration Mercure Buda Hotel
6:00PM	Reception Mercure Buda Hotel
	Welcome from sponsoring organizations *Chair*: Michael Clegg, University of California, Irvine

Monday, March 31

Mercure Buda Hotel

8:00AM	Registration
9:00 – 10:45AM	**Plenary 1. Introduction to the Forum** *Chair*: Roderick Flower, William Harvey Research Institute, Queen Mary University of London • Introductory remarks: Forum Goals, Format, and Logistics

Opening Panel: Framing the Issue—Web of Prevention, Web of Opportunities

- Web of Prevention: Robin Coupland, International Committee of the Red Cross
- Opportunities with International Organizations
  Ottorino Cosivi, World Health Organization
  Alexandre Bartsev, Organisation for Economic Co-operation and Development

Q&A

10:45 – 11:15AM	Coffee break
11:15 – 12:45PM	**Plenary 2. Emerging Life Science and Technology: Challenges and Opportunities for Biosecurity** *Chair*: Indira Nath, LEPRA-Blue Peter Research Center Presentations: • Jason Chin, Cambridge University • Jörg Stelling, Swiss Federal Institute of Technology (ETH) Zurich • Jane Calvert, Edinburgh University Q&A
12:45 – 1:45PM	Lunch
1:45 – 3:00PM	**Plenary 3. Introduction of the Breakout Sessions: Topics and Tasks** *Chair*: Sergio Pastrana, Cuban Academy of Sciences Presentations: • Building a "Culture of Responsibility": Leiv Sydnes, University of Bergen • Identifying Standards for Research Oversight: David Franz, Midwest Research Institute • Providing Science Advice to Governments and International Organizations and Enhancing the Role of the Science Community: Angelo Azzi, Tufts University Q&A

3:00 – 5:30PM **Breakout Session 1**

	Culture of responsibility	Research oversight	Science advising
Chair:	Leiv Sydnes	David Franz	Angelo Azzi
Rapporteur:	Alastair Hay	Neil Davison	Ralf Trapp

Each group will be asked to address a set of specific questions and tasks. Short presentations will begin the discussions.

Presentations during Breakout Session 1:

Culture of responsibility	Research oversight	Science advising
Katsuhisa Furukawa	Elisa Harris	Lucy Horeau
Iris Hunger	David Friedman	Mohamed Sharif
Chandré Gould	David Franz	Ralf Trapp
Animesh Roul		Brief remarks by Piers Millett and Ottorino Cosivi

Followed by general discussions in the sessions

4:00 – 4:30PM Coffee break (taken during breakout session)

5:30PM Breakout Session ends

6:00PM Bus leaves for reception

6:30PM Reception
Institute of Musicology of the Hungarian Academy of Sciences, Haydn Room

Welcome: Norbert Kroó, Vice President, Hungarian Academy of Sciences

Tuesday, April 1

Mercure Buda Hotel

9:00 – 10:30AM **Plenary 4. Awareness about and Attitudes towards Biosecurity**
Chair: Gail Cassell, Eli Lilly and Company

Presentations:
- IAP Statement and CAS activities: Li Huang, Chinese Academy of Sciences
- NAS/AAAS survey: Ronald Atlas, University of Louisville
- Biosecurity workshops: Brian Rappert, University of Exeter and Malcolm Dando, University of Bradford

Q&A

10:30 – 11:00AM Coffee break

11:00 – 1:00PM **Breakout Session 2**

	Culture of responsibility	**Research oversight**	**Science advising**
Chair:	Leiv Sydnes	David Franz	Angelo Azzi
Rapporteur:	Alastair Hay	Neil Davison	Ralf Trapp

Presentations during Breakout Session 2:

Culture of responsibility	**Research oversight**	**Science advising**
Alastair Hay	Ruth Berkelman	Patrick Rubaihayo
Gerald Epstein	David Carr	Koos van der Bruggen

Followed by general discussions in the sessions

1:00 – 2:00PM Lunch

2:00 – 3:30PM Breakout Session 3

	Culture of responsibility	Research oversight	Science advising
Chair:	Leiv Sydnes	David Franz	Angelo Azzi
Rapporteur:	Alastair Hay	Neil Davison	Ralf Trapp

3:30 – 4:00PM Coffee break

4:00 – 5:00PM **Plenary 5. The 2008 BWC Intersessional Meetings**
Chair: Andrzej Górski, Polish Academy of Sciences

Presentation: Ambassador Georgi Avramchev, Permanent Mission of the Republic of Macedonia to the United Nations Office at Geneva; Chair of the 2008 Meetings of the Biological Weapons Convention

Q&A

6:30PM Bus leaves for conference dinner

7:00PM Conference dinner
Kápátia Restaurant

Wednesday, April 2

Mercure Buda Hotel

9:00 – 10:30 AM **Plenary 6. Summary of Breakout Sessions #1, #2 and #3**
Chair: Hernan Chaimovich, Universidade de São Paulo and Academia Brasileira de Ciências

Reports from the rapporteurs of breakout sessions #1, #2 and #3

Discussion

10:30 – 11:00AM Coffee break

11:00 – 12:30PM	**Plenary 7. Summary of the Meeting** *Chair*: Michael Clegg, University of California, Irvine Discussion of lessons/next steps Suggestions for the workshop report
12:30PM	Meeting adjourns/Informal Lunch

PARTICIPANT LIST

Ronald Atlas
United States
University of Louisville

Georgi Avramchev
Macedonia
Permanent Mission of the Republic of Macedonia to the United Nations Office at Geneva

Angelo Azzi
Switzerland
Tufts University
International Union of Biochemistry and Molecular Biology

Martin Iain Bahl
Denmark
Centre for Biological Defence

Alexandre Bartsev
France
Organisation for Economic Co-operation and Development

Ruth Berkelman
United States
Emory University

Katherine Bowman
United States
The National Academies

Jane Calvert
United Kingdom
Edinburgh University

David Carr
United Kingdom
Wellcome Trust

Gail Cassell
United States
Eli Lilly and Company

Hernan Chaimovich
Brazil
Universidade de São Paulo
Academia Brasileira de Ciências

George Chakhava
Republic of Georgia
Tbilisi State Medical University

Kameswara Rao Chavali
India
Foundation for Biotechnology Awareness and Education

Dongli Chen
China
China Arms Control and Disarmament Association

Jason Chin
United Kingdom
Cambridge University

Michael Clegg
United States
University of California, Irvine
The National Academies

Peter Clevestig
Sweden
Stockholm International Peace Research Institute

Ottorino Cosivi
World Health Organization

Robin Coupland
Switzerland
International Committee of the Red Cross

John Crowley
Division of Ethics of Science and Technology
United Nations Educational, Scientific and Cultural Organization

Malcolm Dando
United Kingdom
Bradford University

Neil Davison
United Kingdom
The Royal Society

Gerald Epstein
United States
Center for Strategic and International Studies

Roderick Flower
United Kingdom
William Harvey Research Institute, Queen Mary, University of London

David R. Franz
United States
Midwest Research Institute

David Friedman
Israel
Institute for National Security Studies
Israel Academy of Sciences and Humanities

Katsuhisa Furukawa
Japan
Research Institute of Science and Technology for Society
Japan Science and Technology Agency

Jennifer Gaudioso
United States
Sandia National Laboratories

Andrzej Górski
Poland
Medical University of Warsaw
Polish Academy of Sciences

Chandré Gould
South Africa
Institute for Security Studies

Katalin Hajos
Hungary
Hungarian Academy of Sciences

Elisa Harris
United States
University of Maryland

Alastair Hay
United Kingdom
University of Leeds

Erik Heegaard
Denmark
Centre for Biological Defence

Lucy Hoareau
Seychelles
Division of Basic and Engineering Sciences
United Nations Educational, Scientific and Cultural Organization

Li Huang
China
Chinese Academy of Sciences

Kathrin Humphrey
Germany
The National Academies

Iris Hunger
Germany
Research Group for Biological Arms Control

Jo Husbands
United States
The National Academies

Richard Johnson
United States
Global Helix LLC

Ferenc Jordán
Hungary
Collegium Budapest (Institute for Advanced Study)
Hungarian Academy of Sciences

Barry Kellman
United States
DePaul University

Serhiy Komisarenko
Ukraine
National Academy of Sciences of Ukraine
Commission on Biosafety

Henri Korn
France
French Academy of Sciences

Gabriele Kraatz-Wadsack
Germany
Office for Disarmament Affairs
United Nations

Norbert Kroó
Hungary
Hungarian Academy of Sciences

Filippa Lentzos
Norway
London School of Economics

Irma Makalinao
Philippines
University of the Philippines Manila

Giulio Mancini
Italy
Landau Network-Centro Volta

Lorna Miller
United Kingdom
Defence Science and Technology Laboratory

Piers Millett
United Kingdom
Office for Disarmament Affairs
United Nations

Sospeter Muhongo
South Africa
Regional Office for Africa
International Council for Science

Paul Nampala
Uganda
Uganda National Academy of Sciences

Indira Nath
India
LEPRA-Blue Peter Research Center

Stuart Nightingale
United States
Consultant, National Institutes of Health

Kathryn Nixdorff
United States/Germany
University of Darmstadt

Gabriel Ogunmola
Nigeria
Nigerian Academy of Sciences

Paula Olsiewski
United States
Alfred P. Sloan Foundation

Sergio Pastrana
Cuba
Cuban Academy of Sciences

Alan Pearson
United States
Center for Arms Control and Non-Proliferation

Janos Pusztai
Hungary
Hungarian Academy of Sciences

Simon Takalani Rambau
South Africa
Academy of Sciences of South Africa

Brian Rappert
United Kingdom
University of Exeter

Cris dos Remedios
Australia
Bosch Institute
International Union of Pure and Applied Biophysics

Khalid Riffi Temsamani
Morocco
Ministry of Higher Education & Scientific Research

Animesh Roul
India
Society for the Study of Peace and Conflict

Lajos Rózsa
Hungary
Hungarian Academy of Sciences

Patrick Rubaihayo
Uganda
Makerere University

Ben Rusek
United States
The National Academies

Barbara Schaal
United States
Washington University in St. Louis
The National Academies

Glenn Schweitzer
United States
The National Academies

Yiming Shao
China
Chinese Center for Disease Control and Prevention

Mohamed Mansour Sharif
Libya
National Permanent Committee
for Bioethics and Biosafety

John Sislin
United States
The National Academies

Daniel Sordelli
Argentina
University of Buenos Aires
International Union of
Microbiological Societies

Nina Steenhard
Denmark
Centre for Biological Defence

Jörg Stelling
Switzerland
Swiss Federal Institute of
Technology (ETH)

T.S. Saraswathy Subramaniam
Malaysia
Institute for Medical Research
Ministry of Health

Leiv K. Sydnes
Norway
University of Bergen
International Union of Pure and
Applied Chemistry

Judit Szász
Hungary
Hungarian Academy of Sciences

Terence Taylor
United Kingdom
International Council for the Life
Sciences

Ralf Trapp
Germany
Independent Consultant, France

Koos van der Bruggen
The Netherlands
Royal Netherlands Academy of
Arts and Sciences

Carrie Wolinetz
United States
Federation of American Societies
for Experimental Biology

El Majid Zayer
Morocco
Ministry of Higher Education &
Scientific Research

C

Examples of Projects and Initiatives

RONALD ATLAS UNIVERSITY OF LOUISVILLE, UNITED STATES

Ronald Atlas is co-director of the Center for Health Hazards Preparedness at the University of Louisville in Kentucky. The Center provides training in responding to disasters, including infection control in the event of bioterrorism and medical and public health responses to pandemics. He is former President of the American Society for Microbiology (ASM) and currently is co-chair of the ASM Biodefense Committee. He also chairs the Wellcome Trust Pathogens, Immunology and Population Health Strategy Committee. He is chairing a National Academy of Sciences-AAAS survey project aimed at assessing awareness of and reactions to the dual use dilemma among AAAS members in the life sciences.

MARTIN IAIN BAHL, ERIK HEEGAARD, NINA STEENHARD NATIONAL CENTRE FOR BIOLOGICAL DEFENCE, DENMARK

Summary of Activities at the National Centre for Biological Defence, Denmark

The National Centre for Biological Defence (NCBD) coordinates all activities regarding surveillance of biological weapons (BW) and bioter-

NOTE: This appendix contains material provided by participants in the 2nd International Forum and has been edited only to provide a common format and editorial style.

rorism at Statens Serum Institut. The center constitutes the point of contact for both national and international BW alarms, requests and sample analysis. NCBD houses the national preparedness operation unit, performs and develops assays for sample analysis, and is engaged in several biodefense research projects as well as intelligence work. Furthermore, the center is developing a biosecurity program.

NCBD participates actively in the Australia Group (AG) and has lately been especially involved in discussions related to the misuse potential of synthetic biology i.e., the de novo synthesis of genes or even organisms from chemically synthesised oligonucleotides.

Tasks of the NCBD, Denmark:

- Operations (preparedness) (including analysis of samples)
- Development and testing of assays
- Biodefence research (European Union projects and national projects)
- Development of a Biosecurity programme for Denmark
- International BW work (i.e., EU, AG, Biological and Toxin Weapons Convention (BWC), UN Security Council Resolution (UNSCR) 1540)

Biosecurity

Biosecurity legislation in Denmark is expected to be passed by March 2008 and the centre therefore has many activities within the area of biosecurity research and development (R&D).

Our activities within this area have focused on developing objective risk models for assessing the BW potential of various pathogens. We are currently working on an objective risk assessment model for sensitive equipment, technologies and know-how. Furthermore, we are outlining an implementation plan for biosecurity, including awareness activities and codes of conduct.

ALEXANDRE BARTSEV
OECD, FRANCE

OECD Best Practice Guidelines on Biosecurity

Innovations derived from research on pathogenic microorganisms promise astounding benefits in health, agriculture and other domains of economic activity. The tremendous advances in biology, biotechnology, genomics, proteomics, synthetic biology and bioinformatics in recent years are almost certain to lead to improved health and well-being. Some such biological resources employed in (R&D) for diagnostics, vaccines and therapeutics, however, possess capacity for dual use; they may be misused to develop biological weapons. Research facilities entrusted with possession of such dual use materials have a responsibility to comply with biosecurity measures that are designed to prevent loss or theft and thereby reduce the probability of a bioterrorist attack.

The OECD has provided a forum for its member countries to engage in a dialogue of international cooperation with a view to produce best practice that helps put in place biosecurity measures for Biological Resource Centres (BRCs), which are repositories and providers of high quality biological materials required for R&D and production in various areas of biotechnology.

Some BRCs might handle and exchange hazardous biological materials that have a potential for so-called dual use. Society confers trust in BRCs as custodians of such materials, demanding that responsibility be taken for their safe keeping. In this context, culture collections have long recognized the duties of implementing proper containment procedures for hazardous biological material to safeguard workers against accidental exposure and acting in accordance with legislation on export controls and transport safety measures. More recently, the menace of bioterrorism has added a new dimension to the responsibilities inherent in operating culture collections, namely ensuring security of biological materials with "dual use" potential.

One of the principal challenges in addressing the issues of biosecurity is to find a balance between biosecurity measures that might be applied to BRCs or other research facilities and the access to hazardous biological materials that forms the base for delivering biotechnology innovations.

To qualify the intricacies of such balance, in 2007 the OECD delivered the *Best Practice Guidelines on Biosecurity for BRCs*, which are intended to ensure security of all types of biological materials held by BRCs (e.g., microorganism- and human-derived) in proportion to the risk they present, and thereby marginalize any obstacle that BRCs might face in carrying out their usual operations.

JANE CALVERT
UNIVERSITY OF EDINBURGH, UNITED KINGDOM

Systems Biology

Jane Calvert has been working for the last two years on the social dimensions of the new field of systems biology. She has been interviewing systems biologists, and has spent time at three systems biology laboratories. Dr. Calvert is planning to continue this work at the systems biology centers in Edinburgh and the Imperial College London. Both of these centers also have interests in synthetic biology.

Her main areas of interest in systems biology are:

- the epistemic aspirations of the field
- interdisciplinarity and disciplinary identity

Publication: O'Malley, M.A., J. Calvert, and J. Dupre. 2007. The socioethical study of systems biology. *American Journal of Bioethics* 7(4):67-78.

Synthetic Biology

Calvert's interests in synthetic biology have grown out of her interests in systems biology. She is a member of the UK's Royal Academy of Engineering working party on synthetic biology, and she is also part of two synthetic biology research networks, which bring together natural and social scientists across the United Kingdom.

Calvert's areas of interest in synthetic biology are:

- the relationship between systems biology and synthetic biology
- the relationship between biology and engineering in synthetic biology
- the treatment of complexity and its necessity for living systems
- modularity and open source in synthetic biology
- understandings of "nature" in synthetic biology
- the role of social scientists in synthetic biology

Publication: O'Malley, M., A. Powell, J. Davies, and J. Calvert. 2008. Knowledge-making distinctions in synthetic biology. *BioEssays* 30(1):57-65.

Intellectual Property in the Emerging Life Sciences

Calvert has been working for several years on intellectual property (IP) issues in genomics and genetics and she is interested in pursuing those issues further relation to both systems biology and synthetic biology. She is interested in attempts to patent emergent biological systems and in the applicability of open source principles to the biosciences.

Publication: Calvert, J. 2007. Patenting genomic objects: genes, genomes, function and information. *Science as Culture* 16:207-223.

Translational Research

Dr. Calvert is also interested in the category of "translational research" and asking exactly what this means in policy and scientific terms.

DAVID CARR
THE WELLCOME TRUST, UNITED KINGDOM

The Wellcome Trust is the largest charity in the United Kingdom. It funds innovative biomedical research, in the United Kingdom and internationally, spending around £650 million each year to support the brightest scientists with the best ideas. The Wellcome Trust supports public debate about biomedical research and its impact on health and well-being.

Over recent years, the Wellcome Trust has contributed actively to policy discussions at the UK and international level on addressing risks that life sciences research could be misused for terrorist purposes. We published a position statement on "Bioterrorism and Biomedical Research" in November 2003, which sets out our position on these issues.

In September 2005, the Biotechnology and Biological Sciences Research Council, the Medical Research Council and the Wellcome Trust published a joint policy statement on managing risks of misuse associated with grant funding activities. This statement identified a series of agreed actions that the three organizations have implemented to raise awareness and to help ensure that any risks of misuse associated with research proposals are considered at the grant application stage. We have introduced a standard question on application forms, and ask both our expert referees and our funding committees to consider any risks of misuse associated with the proposals they review.

Further information on our policy work in this area can be found at: *www.wellcome.ac.uk/About-us/Policy/index.htm*.

GEORGE CHAKHAVA
TBILISI STATE MEDICAL UNIVERSITY, REPUBLIC OF GEORGIA

My group focuses on biosecurity issues and national policy as it relates to health and biological sciences. These two areas have melded together in a number of ways since 2006, after avian flu attacks. First, there was a dramatic increase in research on bioterrorism threat agents including anthrax, tularemia, and others. One of the main topics of our interests are also neuroinfections caused by Herpes and Bunya viruses, slow viruses and other interested agents. With this increase came the frightening fact that we have also dramatically increased the number of scientists who have access to and the knowledge of how to handle these agents. Second, what we have not seen is a serious commitment to increasing our nation's public health infrastructure to handle emergencies, including the threat of a pandemic outbreak of influenza. This is absolutely essential, not just for the nation's national security as it pertains to bioterrorism, but for all public health emergencies.

We seek contacts with other universities, societies and institutions to collaborate on joint projects in this field: building a "culture of responsibility" (education and awareness raising, codes of conduct, etc.).

DONGLI CHEN
CHINA ARMS CONTROL AND DISARMAMENT ASSOCIATION, CHINA

The Biosecurity Program in the China Arms Control and Disarmament Association has joint projects with Beijing STS Advisory Center. Current projects include:

- Policy study on strengthening the Biological Weapons Convention;
- Research and dissemination of international and national policies on biological non-proliferation and export control;
- Impact of bioterrorism on bio-arms control and biosecurity;
- Training and education on biosecurity and dual use issues of biotechnology. The project aims to improve awareness of officials, scientists, students and other people from government; medical institutions; research institutions; universities; and industry. This is the emphasis of our current activities.

PETER CLEVESTIG
STOCKHOLM INTERNATIONAL PEACE RESEARCH INSTITUTE (SIPRI), SWEDEN

Dr. Peter Clevestig (Sweden) is a researcher in the Chemical and Biological Warfare Programme of the SIPRI Nonproliferation and Export Controls Project. He is studying the role and responsibility of the Swedish biomedical research community in preventing acts of mass-impact terrorism (funded by the Swedish Emergency Management Agency). The main objective of his project is to raise awareness of biosecurity issues in life science research at academic institutions. He is also developing documentation on biosecurity for use by researchers, heads of laboratories and laboratory management. An additional goal is to review how dual use research of concern is reviewed and assessed from initial conception through to final publication. Dr. Clevestig also has interests in the developing field of microbial forensics in investigating bio-related terrorism and crimes, as well as how emerging fields within life science research are considered from the perspective of dual use and biosecurity. A virologist, Dr. Clevestig holds a doctorate in Infection Biology through his work on HIV-1 in vertical transmission from the Department of Microbiology, Tumor, and Cell Biology at the Karolinska Institute. He also holds a B.Sc. in biomedical laboratory science and B.M.Sc. in biomedical laboratory science. Before joining SIPRI, Dr. Clevestig was administrator of the Karolinska Institute Biosafety Committee and has been an active member of the Nordic Biosafety Network.

OTTORINO COSIVI AND EMMANUELLE TUERLINGS BIORISK REDUCTION FOR DANGEROUS PATHOGENS TEAM, DEPARTMENT OF EPIDEMIC AND PANDEMIC ALERT AND RESPONSE, WORLD HEALTH ORGANIZATION (WHO)

Life Science Research and Development for Global Health Security

The overall goal of the project is to raise awareness and provide information and guidance to WHO Member States on the possible options for risk management to address dual use life science R&D. It underlines the importance of carrying out life science R&D for improving public health and, at the same time, highlights the necessity of understanding that access to, and research on, any type of dangerous agent or new agents may pose risks to public health and raise ethical and security concerns. It therefore aims at involving the public health community on this issue because poorly designed risk management measures will have implications for public health.

The issue is a cross-cutting one—it involves those working with dangerous pathogens but also those working on health research policy, collaboration and support, global health security and ethics. Hence our partnership with WHO departments and external experts that reflect such expertise.

The project started in July 2004 with a grant from the Alfred P. Sloan Foundation. The following phase (2005-2006) of the project was financially supported by the Sloan Foundation and the Ford Foundation. A third phase started (2007-2009) with the support of the Sloan Foundation. Others have expressed interest in financially supporting the project.

Main Achievements

- Establishment of an international network of experts on this subject and in-house collaboration with other WHO programs.

• Publication of working paper identifying the issues from a public health perspective (2005).[1]

• Meeting of a Scientific Working Group to provide guidance on the project activities and publication of the meeting report (October 2006).[2]

• WHO co-sponsorship with the U.S. Government of the international meeting hosted by the U.S. National Science Advisory Board for Biosecurity (NSABB) "International Roundtable on Dual Use Life Sciences Research," Bethesda, MD, February 24-27, 2007.

• Organization and coordination of an online consultation (questionnaire posted on WHO Web site) to receive feedback on the project activities (June-September 2007).[3]

• Organization of a regional workshop on "Research Policy and Management of Risks in Life Science Research for Global Health Security," Bangkok, Thailand, December 10-12, 2007 (in collaboration with our WHO Regional Offices for South-East Asia and for the Western Pacific and support from WHO departments on Ethics, Equity, Trade and Human Rights and Research Policy Cooperation).[4]

• Outreach activities to raise awareness about the project included publications[5] and contributions to more than 30 international meetings and workshops. Technical support was also provided through the col-

---

[1]WHO (World Health Organization). 2005. Life Science Research: Opportunities and Risks for Public Health. Geneva: World Health Organization. WHO/CDS/CSR/LYO/2005.20. Available at: *http//www.who.int/csr/resources/publications/deliberate/WHO_CDS_CSR_LYO_2005_20/*en/index.html.

[2]WHO (World Health Organization). 2007. Scientific Working Group on Life Science Research and Global Health Security: Report of the First Meeting. Geneva: WHO. WHO/CDS/EPR/2007.4. Available at: *http://www.who.int/csr/resources/publications/deliberate/WHO_CDS_EPR_2007_4.*

[3]Online consultation: Scientific working group report feedback form. Available at: *http://www.who.int/csr/bioriskreduction/lifescience_project/en/index.html.* Accessed on December 11, 2008.

[4]The report of the meeting is being drafted.

[5]Reis, A. and E. Tuerlings. 2007. Bioethics and Health Security: The use and misuse of results of life science research. Abstract submitted for the 5th World Conference on Bioethics, Gijón, May 21-25; Tuerlings, E. 2007. Reflections—Governing dual use life science research: Opportunities and risks for public health. In *A Web of Prevention: The Life Sciences, Biological Weapons and the Future Governance of Research*, B. Rappert and C. McLeish, eds. London: Earthscan; Tuerlings, E., and C. McLeish. 2004. Is risk assessment a useful method to govern dual use research? Discussion Paper. 21st Pugwash CBW Workshop: The BWC New Process and the Sixth Review Conference, Geneva, Switzerland, December 4-5; WHO (World Health Organization). 2005. Life Science Research: Opportunities and Risks for Public Health. WHO/CDS/CSR/LYO/2005.20. Geneva: WHO. Available at: *http//www.who.int/csr/resources/publications/deliberate/WHO_CDS_CSR_LYO_2005_20/*en/index.html; WHO (World Health Organization). 2007. Scientific Working Group on Life Science Research and Global Health Security: Report of the First Meeting. Geneva: WHO. Available at: *http://www.who.int/csr/resources/publications/deliberate/WHO_CDS_EPR_2007_4.*

laboration with WHO biosafety and laboratory biosecurity workshops in Iran (October 2006) and Kenya (May 2007).

Forthcoming Activities (2008-2009)

In collaboration with the scientific working group and other WHO departments, the project is now developing a draft guidance document that will complement the two previous project publications. The document will provide guidance on the process to assess national needs and capacities (i.e., how to evaluate needs and capacities to address such risks) and will provide a framework of possible options to manage the risks from a public health perspective (i.e., options will include biosafety and laboratory biosecurity, research policy, and ethical frameworks). The project will also develop technical materials to provide training. This will be done in collaboration with external partners.

To develop the draft guidance and the training, the project is expected to hold two meetings. One meeting will be to review existing risk management practices on the risks posed by life science research and inform the guidance document development. The other meeting, the second meeting of the scientific working group, will be to review the progress of the project, including the final draft guidance and other materials.

Additional outreach activities will be done through non-WHO publications and through contributions to international meetings. For instance, the project will continue its collaboration with the WHO project on Biosafety and Laboratory Biosecurity and will contribute to their regional workshops. Similarly, it will contribute to relevant meetings organized by national academies of sciences and other external partners.

ROBIN COUPLAND
INTERNATIONAL COMMITTEE OF THE RED CROSS (ICRC)

The "Web of Prevention"—A Call for Synergy of Action to Prevent Poisoning and Deliberate Spread of Infectious Disease

The International Committee of the Red Cross launched a Public Appeal in September 2002 on "Biotechnology, Weapons and Humanity." The Appeal carried three messages: first, it drew attention to potential risks brought by advances in life sciences and biotechnology; second, it underscored the legal rules—both national and international—which might apply to poisoning and deliberate spread of infectious disease; and third, it identified responsibilities of both governments and the scientific community to ensure that such advances are used only for the benefit of humanity.

The possible measures to reduce the risk of poisoning and deliberate spread of disease lie in multiple domains, e.g., disease surveillance, criminal law, public health preparedness, international law, codes of conduct, education, etc. Each such measure is necessary but not, in itself, sufficient to reduce this risk. This means that all preventive measures work to enhance each other that is, there is a synergy of action or "Web of Prevention." This is a base concept of the ICRC's initiative. The Web of Prevention makes obvious the links between different agencies working on issues related to biological weapons or chemical weapons, for example, police, scientists, nongovernmental organizations and diplomats. It also discourages compartmentalized thinking and action by different disciplines. Codes of conduct and education within the scientific community only make sense if seen as part of the Web of Prevention.

JOHN CROWLEY
UNITED NATIONS EDUCATIONAL, SCIENTIFIC AND CULTURAL ORGANIZATION (UNESCO)

John Crowley is Chief of Section responsible for science and technology in the Division of Ethics of Science and Technology of UNESCO's Sector for Social and Human Sciences (SHS). He is also editor of the *International Social Science Journal*. He was previously (2005-2007) responsible for SHS information, communication and publications. Before joining UNESCO in 2003, Dr. Crowley worked as an economist in the oil industry (1988-1995) and as an academic political scientist at the French National Political Science Foundation (1995-2002). His research interests cover a number of areas in political theory and comparative politics including, in the areas relevant to his UNESCO responsibilities, environmental ethics and political technologies of securitization.

The section of which Dr. Crowley is head is responsible for three components of UNESCO's programme in the ethics of science and technology:

- Science ethics, including in particular international and interdisciplinary cooperation on the development of codes of conduct for scientific activity, building on the 1974 UNESCO Recommendation on the Status of Scientific Researchers and on the outcomes of the 1999 World Science Conference;
- Environmental ethics, with particular current emphasis on climate change;
- Ethical challenges relating to emerging technologies, including in particular nanotechnologies, new information technologies and issues of biocontrol.

The section also provides the Secretariat for the World Commission for the Ethics of Scientific Knowledge and Technology (COMEST), a body of independent experts established by UNESCO to advise the Director General on issues requiring ethical consideration and, where appropriate, the development of new mechanisms or instruments. In 2008-2009, the work of COMEST, which will next meet in extraordinary session in Paris in November 2008, will focus on science ethics and on climate change.

MALCOLM DANDO
BRADFORD UNIVERSITY, UNITED KINGDOM

In cooperation with Alex Kelle and Kathryn Nixdorff, Malcolm Dando is developing the work they did on "Controlling Biochemical Weapons," looking specifically at the potential misuse of neuroscience.

With Brian Rappert he is working on awareness raising and education for life scientists using the interactive seminar that Brian designed.

NEIL DAVISON
THE ROYAL SOCIETY, UNITED KINGDOM

The Royal Society continues to engage on dual use and biosecurity issues through its membership of the InterAcademy Panel on International Issues (IAP) Biosecurity Working Group and through involvement in UK and international workshops and meetings. The society's standing committee on the Scientific Aspects of International Security maintains a strong interest in this area. Other related areas of interest include export controls, particularly the UK government's new Academic Technology Approval Scheme.

GERALD EPSTEIN
CENTER FOR STRATEGIC AND INTERNATIONAL STUDIES, UNITED STATES

Global Forum on Biorisks (Initiated by the Center for Strategic and International Studies)

The deliberate use of biology for harm can be at once a public health emergency, a crime, a terrorist incident, a disaster, a scientific investigation, and a trade/travel crisis. Moreover, the problem is inherently international, since groups based in one country can acquire resources in a second to attack a third, with the resulting disease spreading to additional countries and its indirect consequences being felt in yet more. Many different professional communities have a role in preventing such incidents, or in managing their consequences. Each of these communities sees one aspect of such an incident, but each may be blind to its many other attributes. To deal with this problem, communities will have to work with one another, including some with which they may be quite unfamiliar, and which may have very different operating procedures, cultures, priorities, and contexts for comparison.

Despite the requirement to work together, no integrated biological risk management governance structure currently exists that is at once comprehensive, international, and multisectoral. Existing governance approaches are not well suited to a problem with such a decentralized set of actors or diversity of perceptions. Intersectoral and international linkage mechanisms are weak and uncoordinated, and no effective way exists to take advantage of potential synergies between the many professional communities that are involved.

The Global Forum on Biorisks constitutes a governance approach that focuses on building linkages and connections among all of the relevant professional communities. The forum will develop a bottom-up, decentralized, adaptive, and interactive mechanism that provides the information and communications necessary to focus attention on biological challenges, facilitate assessment and adaptation among each component of this complex system, and promote interaction among its many moving parts. Implementing this approach involves:

- Working within each professional community to identify modes of operation that remain relevant to managing future biorisks, and those that must be revisited;
- Facilitating understanding within each community of the roles, assumptions, priorities, and values of those other communities with which it must interact;
- Promoting engagement and interaction among communities, sharing information, developing joint projects, and shaping new ways of working together; and
- Providing a driving force to motivate and catalyze action.

The Global Forum on Biorisks is pursuing efforts to accomplish these tasks through two key mechanisms: a highly interactive and customizable Web portal—now under construction—and a series of workshops around the world to introduce this concept and demonstrate the portal. The Web portal will create opportunities for each user to interact, engage, and collaborate. It will introduce social networking tools to the field of biosecurity and host a growing database of information that will be developed and maintained by all of the Web community's participants. No Web site run centrally by any one organization can have the breadth or the currency of one for which all participants take collective responsibility.

DAVID FRIEDMAN
INSTITUTE FOR NATIONAL SECURITY STUDIES AND THE ISRAEL ACADEMY FOR SCIENCES AND HUMANITIES

Oversight of Dual Use Biotechnological Research in Israel

Israeli scientists perform forefront research in the life sciences, biotechnology and biomedicine. They engage in a wide range of projects, using a wide variety of microorganisms, some of them virulent. Based on a heterogeneous research infrastructure, they use all internationally available scientific methods. This research, and related routine work, is conducted in three major sectors: academia (universities and research institutes), hospitals and the biotech industry. Researchers who use virulent microorganisms are obliged, under Israeli law, to follow specific protocols and safety standards, generally those required by such internationally respected groups as the World Health Organization and the U.S. Centers for Disease Control and Prevention. Israel has a comprehensive legislative infrastructure that mandates biological safety (biosafety) procedures. In contrast, Israel lacks a proper legal infrastructure for *biosecurity* (as distinct from biosafety), largely because no one has ever demanded one. Furthermore, since the awareness of its importance is relatively new, it remains minimal. In practical terms, there is a certain amount of overlap between the demands of biosafety and of biosecurity. The existing biosafety procedures do contribute somewhat to biosecurity, but this contribution is far from comprehensive and certainly imperfect. In particular, biosafety rules do not directly address the seepage of dangerous microorganisms and information to hostile elements. The big challenge is to incorporate biosecurity concerns into this system, in particular, to upgrade measures to prevent the leakage of dangerous organisms, information and technologies to terror organizations. To this end the Israel National Security Council and the Israel Academy of Sciences and Humanities initiated a national project and formed a special Steering Committee on Biotechnology Research in an Age of Terrorism (COBRAT) to analyze and report on the current situation and to recommend future action. COBRAT took the above situation as its starting point in seeking more effective and systematic ways to meet biosecurity concerns without compromising academic freedom and creativity. In its final report the Committee formulated specific recommendations to address:

• Changes required in Israel's existing legislative infrastructure.
• Compilation of an updatable list of biological agents and research topics requiring inspection and supervision.
• Establishment of a regime for tracking, supervising and enforcing all areas of biosecurity.
• The need for a national interministerial body or professional committee to guide, monitor and maintain biosecurity.

In pursuing these goals, COBRAT was confronted by several daunting but not atypical facts: (1) no biosecurity legislation exists in Israel, (2) the legislative process, as practiced by the Israeli parliament (Knesset), is long, complicated and uncertain, (3) a response to the bioterror threat cannot wait for long-term solutions. COBRAT's innovative yet practical interim solution to this particular problem may also serve as a useful model for others. As mentioned above, Israel does have a well-developed legal regime that defines biosafety regulations and responsibilities in Israeli governmental, academic and private laboratories. COBRAT, therefore, recommended modifying Israel's biosafety committees and empowering them, by executive order, to undertake responsibility for biosecurity concerns as well. In addition to reducing duplication, disruption, and delay, this scheme avoids many of the sensitivities, suspicions, and conflicts inherent in the regulation of dual use research. The existing biosafety committees are of long standing, they are sensitive to scientific (and personal scientist) concerns, they are well tolerated by the scientific and academic community, and they are unlikely to trigger the hostility and "graft rejection" typical of introducing a "foreign body" into academia. Trust and comfort are intangibles, but their effects are all too real.

Current and Future Activities

• Committee's recommendations approved by Israeli Academy and National Security Council.
• Deliberations in the Israeli Academy and National Security Council on implementation of the committee's recommendations.
• Deliberations in the Ministry of Health on ways to implement recommendations.
• Deliberations in the Israeli Parliament on new legislation based on the recommendations of the committee.
• Dual use and biosecurity sessions in professional associations (e.g., Israeli Microbiological Association).
• Participation in dual use and biosecurity conferences and symposia.
• Preparation of suggestions for programs to raise awareness, consciousness and education in academia.

KATSUHISA FURUKAWA RESEARCH INSTITUTE OF SCIENCE AND TECHNOLOGY FOR SOCIETY, JAPAN

Katsuhisa Furukawa is in charge of the program of Research on Science and Technology for Counter-Terrorism (RISTEX) under coordination with the R&D activities under the Ministry of Education, Culture, Sports, Science and Technology (MEXT) initiative on Science and Technology (S&T) for Safety and Security. The RISTEX team will collect information about current trends and updates about the R&D in this area both in Japan and abroad, analyze them and share the findings with relevant actors in Japan, with the objective to forge a network among researchers, officials and experts as well as between institutions. Through the conduct of this project, the RISTEX project team aims to contribute to making plans and strategies for S&T for counterterrorism as well as their implementation. This project is intended to be one of the efforts to forge a bridge between the scientific community and the national security community both in Japan and abroad.

Within this broad objective, one of the core pillars is to manage dual use aspects of R&D of S&T in order to minimize the risk of misuse of S&T by state actors and nonstate actors. With the objective to develop and institute appropriate governance structure for R&D of S&T at the Japanese universities and research institutions, the RISTEX project team has conducted the following activities:

- Raising awareness among the relevant stakeholders about the importance of the problems of the potential misuse of science and technology, through holding seminars bringing together relevant stakeholders and briefing to officials, experts and political authorities as well as members of the Science Council of Japan.
- Collaborating with the Japanese government in order to develop appropriate measures and policy, including
  - Assisting the development of a guideline for universities and research institutions over the management of R&D activities involving sensitive technologies, within the study group of the Ministry of Economy, Trade, and Industry;
  - Assisting the efforts to develop appropriate measures for biosecurity at universities and research institutions by the Ministry of Education, Science, and Technology;
  - Assisting the efforts of the Ministry of Foreign Affairs to prepare for international cooperation for biosecurity, including the BWC Expert group meetings, the Japan-U.S.-Australia Trilateral Conference's Committee on Counter-Bioterrorism and associated efforts to support capacity building for counter-bioterrorism among ASEAN countries;

– Promoting international cooperation by inviting foreign experts and officials in areas associated with biosecurity and participating international conferences and seminars abroad; and

– Aiming to establish best practice at universities in Japan.

JENNIFER GAUDIOSO
SANDIA NATIONAL LABORATORIES, UNITED STATES

Overview

- Sandia National Laboratories provides innovative, science-based, systems-engineering solutions to the United States' most challenging national security problems.
- Sandia's Global Security Center reduces current and emerging proliferation and terrorism threats by creating sustainable system solutions through international cooperation.
- The International Biological Threat Reduction program, a division of Sandia's Global Security Center, enhances U.S. and international security by reducing biological threats worldwide.

Goals

The three highest goals of Sandia's International Biological Threat Reduction program are:

1. Enhance safety, security, and containment of dangerous biological agents in bioscience facilities.
2. Strengthen capacities to detect and control dangerous biological agents.
3. Improve understanding and mitigation of biological threats.

The International Biological Threat Reduction program advances international threat reduction goals by promoting safe, secure, and responsible use of dangerous biological agents across the globe.

Laboratory Biosafety, Biosecurity, and Biocontainment

- Working with domestic and international bioscience facilities and government agencies to conduct risk, threat, and vulnerability assessments.
- Designing and implementing laboratory biosafety, biosecurity, and biocontainment systems at biomedical and bioscience research facilities across the globe.

- Providing technical assistance to maintain safe and secure transport of dangerous biological agents between facilities.

Infectious Disease Diagnostics and Control

- Designing and implementing modern molecular diagnostics to enhance infectious disease detection and reduce reliance on live, dangerous biological agents.
- Developing outbreak control programs that maintain safe and secure handling of dangerous biological agents in the event of a natural outbreak.

Training and Workshops

- Conducting training to U.S. and international scientists, laboratory managers, and policy makers on the importance of biosecurity, biosafety, biocontainment, and infectious disease diagnostics and control.
- Hosting laboratory biosafety, biosecurity, and biocontainment symposia worldwide, at the request of the international community.

Policy, Regulatory, and Guidelines Support

- Assisting partner countries by reviewing and drafting biosafety, biosecurity, and biocontainment procedures and plans designed to protect dangerous biological agents during handling, research, storage, and transport.
- Helping to develop national and international biosafety, biosecurity, and biocontainment policies, regulations, standards, and guidelines.

Assessments and Analysis

- Conducting assessments and analysis to better understand global biological threats and risks.
- Performing country and regional studies that focus on highly infectious diseases, and the bioscience technologies, expertise, and infrastructure around the world to combat those diseases.
- Developing systematic approaches to prioritize biological threats worldwide, and identifying the best technical solutions to mitigate those threats.
- Writing peer-reviewed publications in the biological threat reduction, nonproliferation, and counterterrorism fields

For more information, visit us online at *http://www.biosecurity.sandia.gov.*

ANDRZEJ GÓRSKI
POLISH ACADEMY OF SCIENCES, POLAND

The misuse of research results for malevolent purposes poses a threat to public health and national security. On February 26-27, 2007, at the invitation of the Director of NIH, Andrzej Górski participated in an international roundtable on dual use life science research held in Bethesda, MD. In his talk presenting the administrative/regulatory activities of the Polish Academy of Science he suggested that an international conference addressing those issues be held in Poland later that year. Its major tasks should be to provide more information about the nature of the dual-use dilemma, to increase the level of awareness of the risks involved, and to discuss possible means of safeguarding research with potential dual use application.

The conference was held under the auspices of UNESCO and the President of the Polish Academy of Sciences (PAS) on November 9-10, 2007 at the Presidential Palace in Warsaw.

The conference presentations are available on the PAS Web site: *http://www.pan.pl/english/index.php?option=com_content&task=view&id=1346&Itemid=287.*

A special issue of *Science & Engineering Ethics* with the conference proceedings will be published later this year.

Furthermore, at the invitation of the European Commission Andrzej Górski presented a talk at the National Ethics Committees Forum in Lubliana (March 2008) entitled "Is Dual Use Adequately Defined and Addressed in Current Research Ethics Guidelines?" In his talk he discussed the definitions, history, examples, and current approaches to dual use, emphasizing the differences between the United States and Europe (regulatory vs. self-regulatory models). The paper is available at the Web site: *http://europa.eu/sinapse/directaccess/NEC/Public-Library/* (page 13).

A manuscript under the same title will be submitted for publication in September.

In conclusion, the conference contributed significantly to the dissemination of knowledge of the dual use dilemma and emphasized the need to further discuss and implement methods to safeguard research with such potential.

CHANDRÉ GOULD
INSTITUTE FOR SECURITY STUDIES, SOUTH AFRICA

Chandré Gould is a senior researcher in the Crime and Justice Programme of the Institute for Security Studies, an African policy research institute. For the past nine years she has been involved in efforts to strengthen the norms against chemical and biological weapons at a national and international level. She is currently involved in the development of an educational module for South African scientists to make them aware of dual use issues and their responsibilities in relation to preventing the misuse of science. She has worked with Dr. Brian Rappert and Dr. Malcolm Dando to organize and present interactive seminars on dual use issues to scientists in South Africa, Kenya, and Uganda.

ELISA HARRIS
CENTER FOR INTERNATIONAL AND SECURITY STUDIES, UNIVERSITY OF MARYLAND, UNITED STATES

Controlling Dangerous Pathogens Project

Dual use biotechnology research poses global challenges that cannot be managed effectively either by traditional arms control or by voluntary self-governance. Legitimate science can create new dangers if a cutting-edge experiment has unexpected results, if findings from research done for benign purposes are misapplied by someone else, or if the line between defensive and offensive biological weapon activities becomes blurred in practice or perception. Moreover, the relevant pathogens, equipment, and knowledge are widely distributed in medical and agricultural research institutions around the world. Efforts to protect against the misuse of biotechnology without impeding beneficial research will require new approaches developed cooperatively by a broad range of stakeholders.

The Controlling Dangerous Pathogens Project brings together leading scientists, security experts, government officials, lawyers, and industry representatives in the United States and in other countries and regions around the world to address the issue of dual use research. As part of this effort, the project has examined the risk that advances in dual use research could lead either inadvertently or deliberately to destructive consequences and explored various national and international mechanisms that could help mitigate this threat.

The Project has developed a detailed proposal for protective oversight of dual use research that would apply comprehensively to all relevant research institutions, whether government, academic or private sector, would rely on mandatory requirements rather than self-governance, and would be global in scope. The project is also seeking to raise awareness on the dual use issue and effective policy responses through a variety of outreach activities, including a series of workshops being held in regions around the world. (Further information on the project can be found at: *http://cissm.umd.edu/projects/pathogens.php/#papers.)*

ALASTAIR HAY
UNIVERSITY OF LEEDS, UNITED KINGDOM

Multiple Uses of Chemicals: Making the Right Choice (A Joint IUPAC/OPCW Project)

Chemical processes determine how we live. From bodily functions through to lifestyle purchases, chemistry is an integral part of who we are and what we aspire to be. But it is not limited to these examples. Many of the major global issues we face will require chemical solutions, be it understanding what happens in the Earth's atmosphere, providing clean water, improving food supplies or discovering new medicines. Chemistry thus has enormous potential to contribute positively to global well-being.

But it has not always been like this. In World War I, chemists and chemical engineers were actively perfecting weapons that relied on the physical and toxic properties of chemicals. Eighty years later the advent of the Chemical Weapons Convention (CWC) changed the rules.

The CWC requires states not to develop or promote chemical weapons. Chemists have a crucial role to play in this process. If the proscriptions of the CWC are to succeed, chemists will have to support them. Many chemists do not know about the CWC. There is a need to inform them about the treaty and about their responsibilities.

To enable this process the Organisation for the Prohibition of Chemical Weapons (OPCW), which is responsible for implementation of the CWC, and the International Union of Pure and Applied Chemistry (IUPAC) have supported a small international working group to develop educational tools to foster debate. Four working papers have been produced which cover multiple uses of chemicals, the CWC, the toxicology of selected chemical warfare agents, and codes of conduct. These papers are available in Arabic, Chinese, French, English, Russian, and Spanish, the working languages of the OPCW.

The working papers are designed for use by teachers of chemistry either in universities or high schools. They provide ample material for a one-hour lecture and much more. With questions at the end to encourage debate and further recommended reading matter there is sufficient material for workshops and projects. Approximately six A4-pages long, the papers have been peer reviewed and tested in workshops in the United Kingdom, Russia, South Korea, and Italy. Participants in the workshops have included chemistry students, teachers, university professors, diplomats and specialists in chemical warfare.

Working papers are currently available in English on the IUPAC Web site (*http://www.iupac.org/multiple-uses-of-chemicals*) in the education section. More papers will be produced and eventually all will be available on the Internet in a form that is useful for teaching.

There is a need to produce material that deals with a range of issues specific to chemistry and that will be attractive and engaging for schoolchildren. This material will deal with ethical issues and direct students to an ethical toolkit that can be adapted for any issue. Work is underway on this programme but is at an early stage.

LI HUANG
CHINESE ACADEMY OF SCIENCES

Li Huang received his Ph.D. in the Department of Microbiology at the University of Guelph, Ontario, Canada, in 1988. He was a postdoctoral fellow in the Department of Biochemistry of the School of Hygiene and Public Health at the Johns Hopkins University, Maryland (1988-1993). He became assistant professor in the Department of Biology at Pomona College, California, in 1993 before joining the faculty in the Institute of Microbiology, Chinese Academy of Sciences, in 1996. He was appointed to full professorship in 1998 and is now director of the State Key Laboratory of Microbial Resources. Dr. Huang's scientific work concerns the isolation and biotechnological exploitation of microorganisms from various environments. He has also been working on biosecurity-related issues since 2001, and is currently a member of the Biosecurity Working Group of the InterAcademy Panel on International Issues.

IRIS HUNGER
RESEARCH GROUP FOR BIOLOGICAL ARMS CONTROL, CARL FRIEDRICH VON WEIZSÄCKER CENTRE FOR NATURAL SCIENCE AND PEACE RESEARCH, UNIVERSITY OF HAMBURG, GERMANY

Iris Hunger heads the Research Group for Biological Arms Control at the Carl Friedrich von Weizsäcker Centre for Natural Science and Peace Research at the University of Hamburg. The aim of the Research Group is to contribute, through innovative research and outreach activities, to the universal prevention of biological weapon development, production, and use. The development of new verification and compliance strategies and concepts and concrete verification measures is the core research area.

The following projects are currently carried out by the Research Group for Biological Arms Control:

- Economic, social and legal aspects of biodefence research.
- Role of non-state actors in promoting nonproliferation and arms control against biological weapons.

• Strengthening the confidence building measure regime under the Biological Weapons Convention.
• Monitoring trade of biological dual use items.
• Harmonizing regulations on research of concern in the life sciences.

Past projects include:

• Controlling weapons by controlling science? The role of natural scientists in bioweapon programs of states.
• International impact of national biosecurity legislation.
• Biosecurity policies at international life science journals.
• Lessons learned from the UNSCOM/UNMOVIC inspections in Iraq.
• New developments in aerosolization technology and implications for biological arms control.

FERENC JORDÁN
COLLEGIUM BUDAPEST, INSTITUTE FOR ADVANCED STUDY AND ANIMAL ECOLOGY RESEARCH GROUP, HUNGARIAN ACADEMY OF SCIENCES, HUNGARIAN NATURAL HISTORY MUSEUM, BUDAPEST, HUNGARY

The Ecology of Security

Ecology is the science of interactions and relationships. Complex ecological systems (like forests or lakes) are complex because they have a large number of parts connected by a large number of interactions. The stability and vulnerability of such systems depends on their topology: critical nodes in interaction networks need to be studied for better understanding dynamics (5). Quantifying key nodes in networks is a central issue in more and more disciplines. Network analysis helps in their identification and characterization. It is the ecological nature of various systems that interacting parts form a whole whose properties depend on the parts but also pose constraints on the parts' behavior. This hierarchical view on many complex systems is the way to predict which parts to defend and which ones to attack in them. It is important to note, however, that structurally reliable networks behave in a persistent way only if their interaction pattern is plastic: for example, prey-switching of predators is a key condition for ensuring stability in model ecosystems. Thus, it may be a key aim to incorporate a desirable extent of flexibility in man-made interactive systems (like governments, armies or other social structures, see (3)). I compare the "network ecology" of ecosystems (6), wasp colonies (1), African nations (2) and the London underground (4). In the last example, it seems to be the case that the terrorist side had performed sophisticated network analysis before the July 7, 2005 attacks since they wanted to bomb at the optimal three stations out of more than 3 million combinations.

References

Bhadra, A., F. Jordán, A. Sumana, S. Deshpande, and R. Gadagkar. In press. A comparative social network analysis of wasp colonies and classrooms: Linking network structure to functioning. *Ecological Complexity*, submitted.

Johnson, D., and F. Jordán. 2007. The web of war: A network analysis of the spread of civil wars in Africa. Paper presented at the annual meeting of the American Political Science Association, Chicago, August 30.

Jordán, F. 2008a. Network analysis: Linking parts to the whole in nature and society. Pp. 240-260 in *Darwinian Security—Perspectives from Ecology and Evolution*, R. Sagarin and T. Taylor, eds. Berkeley and Los Angeles: University of California Press.

Jordán, F. 2008b. Predicting target selection by terrorists: A network analysis of the 2005 London underground attacks. *International Journal for Critical Infrastructures* 4:206-214.

Jordán, F., and I. Scheuring. 2004. Network ecology: Topological constraints on ecosystems dynamics. *Physics of Life Reviews* 1:139-172.

Jordán, F., I. Scheuring, and G. Vida. 2002. Species positions and extinction dynamics in simple food webs. *Journal of Theoretical Biology* 215:441-448.

SERHIY KOMISARENKO
UKRAINIAN COMMISSION ON BIOSAFETY AT THE NATIONAL SECURITY AND DEFENSE COUNCIL, UKRAINE

Main occupation: Academician-Secretary of the Division for Biochemistry, Physiology, and Molecular Biology of the National Academy of Sciences of Ukraine. Other current responsibilities: Chairman of the Commission on Biosafety of the National Security and Defence Council of Ukraine, and Director of Palladin Institute of Biochemistry, Kiev.

Main scientific interests are in immunochemical analysis of proteins and peptides and in interrelation between protein immunochemical structure and their biological function.

S. Komisarenko and his pupils are studying extensively the fibrin polymerization sites and fibrin degradation products with the aid of monoclonal antibodies (and their Fabs) as well as nicotinic acetylcholine receptors and PAR3 on B lymphocytes at different stages of their differentiation.

His team also explores immunity of diphtheria and tuberculosis infection using recombinant proteins and peptides: rA and rB Diphtheria toxin subunits, recombinant surface antigens of various Mycobacteria as well as scFv fragments (single-chain antibodies) against these antigens. They are also using nanoparticles and immunochromatography for the development of rapid diagnostic tools. He is also interested in Biosafety and Biosecurity issues, and in chemical and biological warfare implementation in particular.

GABRIELE KRAATZ-WADSACK
UNITED NATIONS

Development of a Single Comprehensive Database on Biological Incidents

Under the mandate of General Assembly resolution 60/288 entitled "The United Nations Global Counter-Terrorism Strategy," the annexed "Plan of Action" contained elements that aimed to strengthen international capacity under item II.11. One such element was to invite the United Nations system, together with Member States, to develop a single comprehensive database on biological incidents. The database is intended to serve as a platform for receiving detailed technical information on biological incidents worldwide in order to build state capacity to prevent and combat bioterrorism. At present no international comprehensive data resource exists in this area, where data are directly provided by Member States and by this considered "quality-controlled." The biological incident database is currently being developed, together with input from interested Member States and relevant international organizations. Discussions were held on the scope and format of the Biological Incident Database and the provision of actual data. In this context an informal consultative meeting of Governmental experts and representatives from relevant international organizations was convened by the Office for Disarmament Affairs (ODA) in New York from July 24-26, 2007. It is envisaged that the database will be accessed as a secure Web-based application on the Web site of the ODA. Access to any sensitive data will be controlled to address possible proliferation concerns.

Expert Meetings to Update the Technical Guidelines and Procedures to Investigate Alleged Use

The ODA organized two meetings of specialized experts in 2007 in pursuance of the General Assembly resolution 60/288 entitled "The United Nations Global Counter-Terrorism Strategy" of September 8, 2006, which encouraged "the Secretary-General to update the roster of experts and laboratories, as well as the technical guidelines and procedures, available to him for the timely and efficient investigation of alleged use." The roster of experts and laboratories has been updated and currently contains 191 biological experts and 41 chemical experts as well as 59 analytical laboratories. The expert meetings to update the technical guidelines and procedures were attended by experts from 10 Member States, 5 international organizations and 1 nongovernmental organization. In its discussions, the group considered the significant scientific advances that had occurred since the initial drafting of the guidelines and procedures in 1989, particularly in health surveillance, detection and diagnosis. The experts produced a unanimous report which takes into account the substantial developments in the biological area since 1989. In their review and update, the experts also noted the entry into force of the Chemical Weapons Convention and the establishment of the OPCW in 1997 and made reference to the UN-OPCW relationship agreement of 2001. The experts assessed that at this stage only an update of the provisions of the technical appendices of the original guidelines was advisable. The report of the expert group has been transmitted to the Secretary-General.

DAVID MBAH
CAMEROON ACADEMY OF SCIENCES, CAMEROON

Brief Summary of the Cameroon Biosecurity Project

The biosecurity project Cameroon is working on is aimed at building national capacity to safeguard, control, monitor and manage genetically modified organisms and invasive alien species including pathogens for the sustainable management of Cameroon's biodiversity and building capacity to implement the Cartagena Protocol on Biosafety and Cameroon's biosafety legislation. Project components include institutional capacity building for management and control of invasive alien species including pathogens and implementation of the Cartagena Protocol, development of legislative/regulatory instruments (IAS, policy on biotechnology/biosecurity/access and benefit sharing), production of documents, transaction/interpretation and management. The project has already received Global Environmental Facility (UN Environment Programme) approval.

LORNA MILLER
DEFENCE SCIENCE AND TECHNOLOGY LABORATORY, UNITED KINGDOM

Lorna Miller is Senior Biological Advisor/Non-Proliferation at the Defence Science and Technology Laboratory, Porton Down, United Kingdom, providing scientific and technical advice on biological arms control and non-proliferation issues to policy makers and implementers. This includes the role of scientific and technical advisor to the UK delegation to the Biological and Toxin Weapons Convention, and support to UK policy on biological export controls. She has provided the technical lead for UK input to the previous BWC Meetings of Experts. Her current work involves preparation for UK contributions to the 2008 Meeting of Experts topics related to biosafety and biosecurity and to oversight, education, awareness raising and codes of conduct to prevent misuse of advances in bioscience and biotechnology research. She is also involved in providing and coordinating technical support to UK initiatives to provide assistance to other countries in meeting arms control and non-proliferation objectives, particularly with regard to implementation of the BWC; biosafety and biosecurity standards and training; and capacity building in infectious disease control.

SOSPETER MUHONGO
ICSU REGIONAL OFFICE FOR AFRICA, SOUTH AFRICA

Professor Sospeter Muhongo, a Tanzanian, is the founding and the current Regional Director of the International Council for Science (ICSU) Regional Office for Africa. He is the chair of the Science Programme Committee of the UN-proclaimed International Year of Planet Earth (IYPE). Professor Muhongo is the Chairperson of the Steering Committee of the Scientific, Technical and Research Commission of the African Union. He is the chairperson of the UNESCO/International Union of Geological Sciences/International Geological Correlation Programme Scientific Board and Vice President of the Commission of the Geological Map of the World. Professor Muhongo is a Full Professor of Geology, a Chartered Geologist (UK) and the first recipient (2004) of the Robert Shackleton Award for Outstanding Research in the Precambrian Geology of Africa. In 2006, Professor Muhongo won the prestigious National Award for Outstanding Research in S&T in Tanzania. The Geological Society of South Africa conferred the prestigious "Honours Award" upon Professor Muhongo for his meritorious contribution to the Earth Science profession in 2007. He is an Honorary Professor of Geology at the University of Pretoria (South Africa). Professor Muhongo has established excellent and reliable contacts with senior scholars, business executives, politicians and government officials throughout the world. He is actively mentoring young scientists around the world and has developed a special interest in the application of science, research, technology and innovation for the sustainable socioeconomic development of the global society.

PAUL NAMPALA
UGANDA NATIONAL ACADEMY OF
SCIENCES (UNAS@INFOCOM.CO.UG)

Promoting Biosafety and Biosecurity within the Life Sciences

The overall aim of this project is to promote policies and practices that will reduce the likelihood of the inadvertent or deliberate spread of disease stemming from life science research. It is intended to serve as a model for other undertakings on laboratory biosafety and biosecurity for African researchers and others around the world. It will also educate and further develop the skills of media representatives in reporting on issues of biosafety and biosecurity. This project was designed to bring together leading scientific and policy experts in Africa at a workshop held in Kampala, Uganda, March 11-12, 2008, to discuss biosafety and biosecurity and to build capacity within research institutions in East Africa to devise and undertake laboratory biosafety and biosecurity oversight review procedures. Beyond the workshop, the project will provide an opportunity to disseminate emerging educational materials and tools and raise the profile of Africa and African countries in international biosecurity deliberations.

The workshop attracted 75 participants consisting of practicing scientists, biosafety officers, policy officials, media representatives, non-governmental organizations, policy analysts and African and U.S. Academy representatives. The symposium topics presented over the course of the two days included the following, among others:

- Overview of laboratory biosafety and biosecurity—International discussion
- Laboratory biosafety and laboratory biosecurity—East African perspective
- Promoting wider engagement about biosecurity and laboratory biosafety
- Educating the media on communicating issues of dual use research and bioterrorism

The symposium raised more questions than answers and a common understanding was sought to clarify the scope of biosafety and biosecurity. The key issues raised include the following:

- Biosafety is accepted as essential to keep up with the rapid developments in biotechnology.
- In Africa, primary biosecurity risk stems from nature and not laboratory undertakings but there is potential for some infectious agents

to spread either accidentally or deliberately from the laboratory and endanger the public so attention to laboratory biosafety and biosecurity is critical.

- Individual countries need to adapt their existing or new legal policy frameworks to capture aspects biosafety and biosecurity.
- There is need for compliance and enforcement of existing laws and regulations, including biopiracy and intellectual property rights.
- The responsibility of biosafety and biosecurity lies at multiple levels including individual, institutional, and oversight responsibilities at national levels.
- The need to educate and sensitize at all levels is apparent. Similarly, capacity building is critical at all levels. To ensure cost-effective approaches, it is important to share training and educational resource materials. Thus, collaboration and effective partnerships should be encouraged.
- African academies should play a more active role in advising governments on biosafety and biosecurity.

STUART NIGHTINGALE CONSULTANT, NATIONAL INSTITUTES OF HEALTH, OFFICE OF BIOTECHNOLOGY ACTIVITIES, OFFICE OF SCIENCE POLICY, OFFICE OF THE DIRECTOR, U.S. DEPARTMENT OF HEALTH AND HUMAN SERVICES

The National Science Advisory Board for Biosecurity (NSABB)

The NSABB is a critical component of a set of federal initiatives to promote biosecurity in life sciences research. The U.S. government established this advisory body to recommend ways to minimize the risk that information from legitimate life sciences research could be intentionally misused to threaten public health and other aspects of national security. The NSABB consists of 25 nongovernment voting members with a broad range of expertise, including molecular biology, microbiology, infectious diseases, biosafety, public health, veterinary medicine, plant health, national security, biodefense, law enforcement, scientific publishing, and related fields. Representatives from 15 federal agencies and departments are nonvoting members. The NIH Office of the Director administers and manages the board. The NSABB is charged specifically with providing recommendations for the development of:

- A system of institutional and federal oversight that allows for fulfillment of important research objectives while addressing national security concerns;

- Guidelines for the identification and conduct of research that may require special oversight;
- Codes of conduct for scientists and laboratory workers that can be adopted by professional organizations and institutions engaged in life science research;
- Materials and resources to educate the research community about effective biosecurity; and
- Strategies for fostering international engagement on dual use biological research issues.

To date the NSABB has developed two major reports that have been transmitted to the U.S. government for consideration during the policy development process: "Proposed Framework for the Oversight of Dual Use Life Sciences Research: Strategies for Minimizing the Potential Misuse of Research Information" and "Addressing Biosecurity Concerns Related to the Synthesis of Select Agents."

In fulfillment of its charge to recommend strategies for fostering international dialogue on dual use research issues, the NSABB has hosted two International Roundtables, the first co-sponsored by the U.S. government and the World Health Organization. The purpose of the Roundtables is to begin a dialogue with scientists and representatives from interested countries, relevant intergovernmental organizations, and scientific professional associations; hear the perspectives of scientists in other countries; learn about relevant activities; and share the tools developed by the NSABB for the identification, conduct, and review of dual use life sciences research. The summaries of these International Roundtables (February and October 2007) are posted on the NSABB Web site. A third International Roundtable is planned for late 2008 and will facilitate outreach to and feedback from scientists and representatives from additional countries.

For more information, please visit the NSABB Web site: *http://oba.od.nih.gov/biosecurity/.*

KATHRYN NIXDORFF DEPARTMENT OF MICROBIOLOGY AND GENETICS, UNIVERSITY OF DARMSTADT, GERMANY

Life Sciences Revolution and Biochemical Arms Control

Characteristic of the developments in science and technology over the past three decades is the explosive nature of the accumulation of knowledge concerning the mechanisms and functions of biological systems. The revolution in biotechnology is continuing on into the revolution in pharmacology with the emphasis on drug discovery and drug delivery, in which biochemical bioregulators (organic chemical substances that regulate the function of biological systems) and systems biology will be gaining more and more significance for biochemical arms control as time progresses. Bioregulators used in a malign way pose a particular threat in that they can be used to disrupt the balanced operation of interacting physiological systems. An example can be found in the interactions of the neuroendocrine and immunological systems, with their vulnerability to compounded modulation. The interdependence of the reaction pathways of these systems raises the dual use dilemma to a whole new order of complexity. Trying to deal with this complexity in order to exploit the benefits while minimizing the risks is going to be an enormous task in the future.

This constellation of factors raises the question as to whether the Biological and Toxin Weapons Convention, which has no treaty organization and does not contain adequate measures for assuring compliance, is running into the danger of being completely overwhelmed by scientific and technological advances in the future in the sense that States will be reluctant to devote appropriate attention to these developments in all their complexity.

Under Article IV of the BWC States Parties have an obligation to take all "necessary measures" not only to "prohibit" but also to "prevent" the malign misuse of biological materials. While penal legislative measures can contribute significantly toward prohibition of misuse, monitoring of developments in science and technology along with the formulation of biosecurity regulations that take these developments into account are counteractive measures that can help prevent the misuse of biological materials.

In the research project, developments in science and technology will be monitored and an in-depth analysis of several main targets of interacting physiological systems that may be used malignly for offensive military purposes will be provided, along with an analysis of the consequences of modulating these targets with biochemical bioregulators. In addition, new developments in pertinent delivery systems that could be used to direct bioregulators to their targets will be investigated, and the feasibility of their application analyzed. Suggestions will be offered for minimizing the risks posed by these developments.

This is a summary of part of a project being conducted in collaboration with Professor Malcolm Dando of the University of Bradford, United Kingdom, and Dr. Alexander Kelle of the University of Bath, United Kingdom, funded by the German Foundation for Peace Research (*http://www.bundesstiftung-friedensforschung.de*).

ALAN PEARSON
CENTER FOR ARMS CONTROL AND NON-PROLIFERATION, UNITED STATES

Alan Pearson, Ph.D., is the Director of the Biological and Chemical Weapons Control Program at the Center for Arms Control and Non-Proliferation, where he is responsible for coordinating the work of the Scientists Working Group on Biological and Chemical Weapons, monitoring U.S. biodefense activities, and promoting national and international efforts for biological and chemical weapons control. He is currently working on issues of national BWC compliance review and enhancing transparency of biodefense and other dual use life sciences research. He is co-editor of the book *Incapacitating Biochemical Weapons: Promise or Peril?* (Lexington Press, 2007). In 2003-2004, Dr. Pearson was an American Association for the Advancement of Science/Nuclear Threat Initiative science policy fellow at the Department of Homeland Security, where he worked in the Biological and Chemical Countermeasures Portfolio of the Science and Technology Directorate. Dr. Pearson was a Research Fellow at the Harvard University School of Medicine and a postdoctoral Research Fellow of the American Cancer Society. He received his Ph.D. in biology from the Massachusetts Institute of Technology.

SIMON TAKALANI RAMBAU
ACADEMY OF SCIENCES OF SOUTH AFRICA

Simon Takalani Rambau is a National and International Liaison Officer at the Academy of Sciences of South Africa (ASSAf). His main responsibility is to coordinate all ASSAf international activities such as to maintain the bilateral and multilateral engagements with other international science academies and organizations such as the InterAcademy Medical Panel, IAP, Network of African Science Academies, African Academy of Sciences, G8 + 5, IBSA (India/Brazil/South Africa) activities, African Union, NEPAD (New Partnership for Africa's Development), ICSU, the Academy of Sciences of the Developing World (TWAS), and the Southern Africa Development Community in order to advance scientific work in Africa. Other responsibilities include serving as an Executive Secretary for Committee of Heads of Organisation for Research and Technology, coordinating the establishment of ASSAf Standing Committee on Biosafety and Biosecurity as well as facilitating the establishment of a South African Chapter of the World Academy of Young Scientists in Africa. He is currently a Ph.D. student at the University of Pretoria conducting research on disaster education in the informal settlements.

C. KAMESWARA RAO FOUNDATION FOR BIOTECHNOLOGY AWARENESS AND EDUCATION, INDIA

In the developing countries, the level of scientific/technical expertise and infrastructure to face sudden and unforeseen threats to health security is pathetically low. There is hardly any expertise for anticipating risk, its assessment and mitigation. Awareness of preparative and remedial action is totally lacking even among the professional and administrative circles, who should know better. The governments and health personnel are totally unprepared, helpless, and lost in times of crisis. It would take a decade for advancements in life sciences/biotechnology that would make significant difference to pathogen diagnostics, preventive and management protocols in the form of vaccines, antibodies, etc., to reach the developing countries. By that time new developments would have replaced them in the West, leaving the developing countries chasing technological innovation forever.

Vast numbers of the population are extremely vulnerable to disease outbreaks, even from conventional threats, resulting in an enormous human tragedy. This is acutely reflected in the past events in India, such as the Surat plague epidemic, outbreak of epidemics following the earthquake in Gujarat, and several other disasters. The confusion and panic on epidemic outbreaks at the periodical and/or annual religious events at auspicious rivers where millions congregate have repeatedly shown the inadequacy of preparedness of public health authorities, and no lessons were ever learnt from the past nightmares. Risks from new epidemics such as bird flu are looming large with no one in authority being visibly conscious of the imminent threat as reflected by the recent outbreak of bird flu in the State of West Bengal. Dual use technologies are beyond all comprehension and remedy.

In order to prepare the developing countries, even marginally, (a) to foresee and face unexpected events, (b) to avoid wasteful duplication of technical and financial resources, (c) to enable them to adopt new technology sooner rather than later, and (d) to minimize, if not totally avoid, the confusion and panic that prevails in times of natural disasters and epidemic outbreaks that pose a serious risk to health security, the following international organizations need to be set up:

1. An international organization with units in different countries to (a) share and provide state-of-the-art technical know-how, and (b) to coordinate and monitor diagnostic, preventive and remedial action and

2. An international funding body to provide financial support to prevent human tragedy for want of technical know-how and financial resources, both lacking in the developing countries.

The mandate of these organizations would be to facilitate

- Improvements in state and local surveillance infrastructure, such as establishment of biodisaster control centers, to perceive and handle bioterrorism and biowarfare threats;
- Survey major hospitals for supplies of antidotes, drugs, ventilators, personal protective equipment, decontamination capacity, mass-casualty planning and training, isolation rooms for infectious disease, and familiarity of staff with the effects and treatment of biological threats;
- Encourage governmental and private agencies engaged in health and medical R&D to share their information on (a) diseases, (b) diagnostics (c) drugs, (d) personnel, (e) resources, and (f) on the sources of threats to health security and protocols to mitigate threats;
- Convene discussions among the appropriate agencies making them aware of current developments and on the use of investigational products in mass-casualty situations and on acceptable proof of efficacy for products where clinical trials are not ethical or are otherwise impossible;
- Develop incentives for both public and private hospitals to be receiving hospitals, to stockpile antidotes and selected antitoxins and make them available to the first responders, by changing laws if needed;
- To purchase appropriate personal protective equipment and expandable decontamination facilities and train emergency department personnel in their use;
- Provide for state and central training initiatives with a programme to incorporate existing information on threats to health security and their preventive and treatment methods into the manuals and reference libraries of first responders, emergency departments and biotoxin control centers; and
- Intensify Public Health Service efforts to organize and equip Urban Medical Response Teams and Community Response Teams, in high-risk cities and other locations, throughout the country.

These measures would enhance the general ability of governments, public health authorities and the communities, to cope with mass-casualty events.

In view of its expertise, influence and reach, the WHO is best suited to recommend and oversee the implementation of the suggested measures.

BRIAN RAPPERT
UNIVERSITY OF EXETER, UNITED KINGDOM

Raising Awareness of Dual Use Research in the International Life Science Community

A major plank of policy responses in relation to this has been devising educational and awareness training for scientists regarding the "dual use" potential of research—its potential to be used for both beneficent and malevolent purposes. Yet, that overall agreement is belied by the lack of specification about the content and specific aims of such provisions. Should that, for instance, consist of providing information on the history of biological warfare, stimulating generic concerns about the responsibilities of scientists today, alerting researchers to security considerations for their individual consideration, or challenging certain presumptions about the malign potential of research?

First as part of a grant funded by the Economic and Social Research Council and more recently through two Alfred P. Sloan Foundation grants, Malcolm Dando (University of Bradford) and Brian Rappert (University of Exeter) are conducting seminars for practicing researchers. There are two aims to this: first, to inform participants about current biosecurity dual use debates, and second, to generate interactive discussion about the merits of proposed policy responses.

The work has been undertaken in collaboration with a number of individuals: Animesh Roul, Society for the Study of Peace and Conflict, India; Peter Edopu and Chandré Gould of the Institute for Security Studies, South Afrcia; David Friedman at the Institute for National Security Studies, Israel; Katsuhisa Furukawa, Rui Kotani, and Yu Sasaki at the Research Institute of Science and Technology for Society in the Japan Science and Technology Agency; Heide Hackmann, Laura van Veenendaal, and Rudie Trienes of the Royal Netherlands Academy of Arts and Sciences; Mari Linnapuomi of the Finnish Ministry for Foreign Affairs; Serhiy Komisarenko of the National Academy of Sciences of Ukraine; Silvia Cucovaz of the Argentinean National Intelligence School; Paula Austin of Sandia National Laboratories, United States; Christian Enemark of the Centre for International Security Studies at the University of Sydney, Australia; and Thomas Egwang of Med Biotech Laboratories and the Ugandan National Academy of Sciences.

Through this work we are developing a novel research method for engaging with practicing scientists about emerging areas of societal discussion. We have also produced interactive educational material, in part, in collaboration with Marie Chevrier (University of Texas at Dallas).

KHALID RIFFI TEMSAMANI, EL MAJID ZAYER
MINISTRY OF HIGHER EDUCATION & SCIENTIFIC RESEARCH, MOROCCO

Morocco's Current Biosecurity Projects

- Morocco participated with the U.S. National Science Advisory Board on Biosecurity in a roundtable on the dual use of life sciences research in February 2007 in Washington, DC.
- Morocco is a member of the Biosafety and Biosecurity core group for the Middle East and North Africa (MENA) region.
- The core group will meet in Abu Dhabi next March 2008.
- Morocco has been chosen to organize the Second International Conference on Biosafety and Biosecurity in March 2009 (BBIC09).
- Morocco is involved in the development of a regional strategy for the Middle East and North Africa, which would underpin and support national strategies, to enhance biosafety and biosecurity. Human, animal and agricultural sectors are targeted in this strategy.
- Academia, governments and private sectors will be part of the strategy elaboration.
- The core group has decided to establish regional biosafety and biosecurity training centers.
- Morocco has started discussions to put in place a National Science Ethics Commission and a national code of conduct for science and technology.

ANIMESH ROUL
SOCIETY FOR THE STUDY OF PEACE AND CONFLICT, INDIA

Animesh Roul is involved as India coordinator in a Sloan Foundation funded "biosecurity" project under Professors Brian Rappert and Malcolm Dando, referenced above. In India, he is also involved in issues relating to bioterrorism and emerging and reemerging infectious diseases and their sociological impact.

LAJOS RÓZSA
THE HUNGARIAN ACADEMY OF SCIENCES, HUNGARY

A Drug Weapon Research Program (1962-1972) of the Warsaw Pact

Contrary to widespread rumors in the Cold War era, there had been little, if any, evidence in the scientific literature to support the view that the Soviet Union or its Warsaw Pact allies considered the use psychochemical weapons within a military context.

The Hungarian State Archives has recently opened up declassified records of Hungary's State Defence Council's meetings held between 1962 and 1978. Materials submitted to the Council include reports on the coordinative meetings of the Warsaw Pact military medical services. According to these reports, research into possible countermeasures against psychotropic drugs was a research priority assigned to Hungary in 1962. Hungary rejected this task in 1963, but joined the ongoing project again in 1965. Methylamphetamine was produced in Budapest for use as an experimental model of such weapons. Within the context of contemporary Western research, this drug was considered an effective interrogation tool. Similar to the contemporary CIA, Hungary also failed to develop an antidote against it and thus the project terminated fruitlessly in 1972.

These documents serve evidence that a Warsaw Pact forum had, in fact, been considering a psychochemical as a weapon.

BARBARA SCHAAL
WASHINGTON UNIVERSITY, ST. LOUIS, UNITED STATES

Barbara Schaal's group studies the genetics of rice. In collaboration with scientists from Chiang Mai University, they are accessing the potential environmental impact of genetically modified rice. Their work centers on gene flow between cultivated rice and native or weedy rice populations in Southeast Asia and in the United States. Their studies have quantified the levels of gene flow and have shown that gene flow between cultivated rice and its wild ancestor results in the development of a weedy variant of rice. Weedy rice is a serious pest in rice fields and can result in large losses of yield and potentially abandonment of rice fields. The implications for such gene flow and hybridization from genetically modified rice have been inferred by a study of comparative fitness of genetically modified rice, wild rice, and their hybrids.

DANIEL SORDELLI
UNIVERSITY OF BUENOS AIRES, ARGENTINA

Current Projects

Dr. Sordelli's work involves two different fields. As a basic science researcher, he is involved in work aimed at increasing the knowledge on mechanisms of pathogenesis and prevention of staphylococcal human bone infections, with special emphasis in capsule expression and infection chronicity. As president-elect of the International Union of Microbiological Societies (IUMS) and chair of the IUMS Public Policy Committee he is leading the drafting of the Health and Science Diplomacy Initiative (HSDI).

Health and Science Diplomacy Initiative

The impact of scientific advances on the world's population, economic, social and political systems has grown dramatically in the last decade. New emerging and reemerging infectious diseases have taken at the same time a significant toll on many countries and populations. Biotechnology and environmental science issues have also had significant impact. Indeed, recent major disasters such as the tsunami affecting Northern Indonesia and many other countries in the region, or current threats such as the potential avian influenza pandemics or the misuse of microorganisms for terrorist acts are examples of scenarios that require immediate attention by a coordinated and balanced team of experts and politicians.

The emergence of these and other multinational issues in the 1990s has redefined the scientific demands placed on those responsible for international relations. Indeed, the sciences—and the microbiological disciplines especially—now play a new role in countries' evolution and are becoming today a major component of government foreign policy.

At a time of globalization, governmental institutions in both developed and developing countries as well as international organizations are facing major challenges in finding effective ways to utilize state-of-the-art scientific and medical advances that offer opportunities that were unimaginable only a few years ago. As a consequence, policy makers worldwide are under increasing pressure and scrutiny to rapidly determine the validity of the science through an understanding of the concepts and to accurately evaluate the potential advantages and disadvantages of each scientific advance for various societies. In this context, strengthening the role of scientists while educating the policy makers and diplomats has become a critical issue.

The main objective of the HSDI would be to mobilize expertise to enable diplomats and government representatives to participate fully and to make informed decisions on emerging issues where science and health (especially from the microbiology viewpoint) play an important role.

The Initiative would provide succinct briefs on emerging microbiological science and technology issues and would analyze information from international treaties, and protocols and international initiatives and events. It would examine the current and future place of science and health diplomacy as a tool for advancing collective interests, with attention to security and globalization issues.

Proposals for pertinent training are expected for diplomats, scientists and policy makers to assist them in international negotiations (particularly those that take place under the UN and related organizations), with respect to the issues of emerging infections, biological diversity, biotechnology, biosecurity, transfer of technology, trade, industry and sustainable development.

T.S. SARASWATHY SUBRAMANIAM
MINISTRY OF HEALTH, MALAYSIA

Current Research

- Acute Flaccid Paralysis Surveillance in Malaysia (Viral etiology and disease spectrum).
- HIV—Immunogenetic factors influencing disease progression

Laboratory Biosafety and Biosecurity

- Chairperson, National Standards Sub-Committee reviewing Malaysian Standards for Code of Practice for Safety in Microbiology Laboratories, 2008.
- Member, Expert Working Group, EWG-BWC, Ministry of Health Malaysia.
- Implementing programs for laboratory capacity and capability in biosafety and biosecurity at national level, code of ethics, practice or conduct for scientists.
- Working Group preparing Guidelines for contained use of LMOs pursuant to Biosafety Act 2007.
- Secretary/Biosafety Officer, IMR Laboratory Biosafety & Biosecurity Committee.
- Implementing safe practices (documentation, manuals, guidelines) in biosafety at IMR.

TERENCE TAYLOR
INTERNATIONAL COUNCIL FOR THE LIFE SCIENCES (ICLS)

The Global Health and Security Initiative (GHSI)

The Global Health and Security Initiative, a project of the Nuclear Threat Initiative (NTI), is working around the world to address the complex and multifaceted risks posed by biological agents. Infectious disease can emerge from many sources, afflicting humans and animals, or threatening the food and water supply. The biological threat—whether from natural disease epidemics, the intentional use of pathogens for harm, or the inadvertent release from a laboratory mishap—is a real and growing concern and challenges traditional ways of thinking about prevention, deterrence and response. Responding to the full spectrum of biological risks requires new thinking about how to identify and implement enduring solutions. GHSI seeks to raise awareness and leverage direct action through innovative partnerships and creative approaches to reducing these threats. A more secure world demands that gaps in the global public health capacity for rapid detection and response be addressed, as well as strengthening efforts to prevent the development and use of biological weapons.

The two goals of the Initiative are to:

1. Promote the safe and secure practice of the life sciences by safeguarding access to dangerous pathogens and preventing the misuse of technology and information.
2. Improve the global capacity for the prevention of and preparedness for biological threats through enhanced disease surveillance, in particular through early detection and response.

Mission Approach and Funding Philosophy

NTI's Global Health and Security Initiative promotes threat reduction solutions, raises public awareness, and undertakes sustainable direct action projects that demonstrate innovative ways to reduce threats. The majority of the Initiative's awards support operational activities that it has a strong hand in developing. The GHSI will undertake and support projects that:

- Address significant high-risk situations;
- Generate additional funding and leverage action for threat reduction; and
- Promote the core objectives of the Global Health and Security Initiative through sustained engagement.

Projects

Examples of projects currently under way to achieve the GHSI mission include:

- Support for the International Council for the Life Sciences, which is the primary vehicle for GHSI in establishing and empowering standing national and regional networks for promoting best practices, standards and training in biosafety and security.

• Establishing and helping to maintain regional infectious disease surveillance consortia to improve their technical capacity in rapid detection, identification and response to infectious disease outbreaks. The principal groups being supported at present are the Middle East Consortium for Infectious Disease Surveillance and, in cooperation with the Rockefeller Foundation, the Mekong Basin Disease Surveillance Network.
• Support for specific facilities and individuals in Russia aimed at enhancing biological safety and security.

A critical element in these strands is the engagement of all sectors of the life science community including academia, government and the private sector. Of particular importance is seeking and encouraging public/private partnerships to bring novel technical solutions and approaches to help reduce biological risks along the full spectrum from naturally occurring events, through accidents or negligence in laboratories to deliberately induced disease outbreaks.

For more information, please see *www.ghsi.org* and *www.iclscharter.org*.

RALF TRAPP
INDEPENDENT CONSULTANT

Ralph Trapp is an independent consultant on disarmament of chemical and biological weapons. He advises the OPCW on the preparation of the Second CWC Review Conference and acts as legal coordinator of the European Union (EU) joint action in support of the BWC (working through the Biological Weapons Prevention Project in Geneva).

KOOS VAN DER BRUGGEN
ROYAL NETHERLANDS ACADEMY OF ARTS AND SCIENCES

A Code of Conduct for Biosecurity in the Netherlands: An Example to Be Followed?

The Dutch Ministry of Education, Culture and Science asked the Royal Netherlands Academy of Arts and Sciences (KNAW) to provide it with advice and input for a national Biosecurity Code of Conduct for scientists, as recommended by the BWC, which was ratified in 1972. The request arose in part from the KNAW's active contribution to the Statement on Biosecurity issued by the InterAcademy Panel in 2005.

If a code of conduct is to have the intended effect, it must reflect the experience and practice of the relevant actors. It was therefore decided to establish a focus group whose members would make comments and suggestions based on their practical experience as researchers and policy makers. The first step of the project was to conduct a survey of measures already taken by central governments, fellow academies and research institutions in other countries, including the United States and the United Kingdom. A further survey was made of current legislation and existing codes of conduct for biotechnology and microbiology with relevance for biosecurity. The findings of these surveys were used to identify how the adoption of a code of conduct can help to ensure that biosecurity issues are effectively addressed in scientific research. The Dutch Biosecurity Code of Conduct, published in October 2007, is accompanied by an explanatory memorandum and a background review, which were also submitted to the working group and the focus group for comment.

The aim of this code of conduct is to prevent life sciences research or its application from directly or indirectly contributing to the development, production or stockpiling of biological weapons, as described in the BWC, or to any other misuse of biological agents and toxins. Given this aim different target groups can be distinguished, varying from professionals engaged in the performance of biological, biomedical, biotechnological and other life sciences research to funding organizations and authors, editors and publishers of life sciences publications. The Code of Conduct on Biosecurity is intended to make all these groups aware of the potential dual use of the results of biological research and to make them follow some basic principles that can reduce the risks.

How this process of this awareness raising on biosecurity issues can be organized will be elaborated and explained on the basis of the Dutch example. In the international context of the 2nd International Forum on Biosecurity attention will be paid to the question if and how the Dutch Code of Conduct on Biosecurity can be an example to be followed for other countries.

CARRIE WOLINETZ
FEDERATION OF AMERICAN SOCIETIES FOR EXPERIMENTAL BIOLOGY (FASEB), UNITED STATES

FASEB Engagement in Dual Use Research Issues

The Federation of American Societies for Experimental Biology comprises 21 scientific societies representing more than 80,000 biomedical researchers. FASEB's mission is to advance biological science through collaborative advocacy for research policies that promote scientific progress and education and lead to improvements in human health. Our societies' members represent both basic and clinical researchers, primarily based in the United States but with a rapidly growing international membership as well.

Dual Use/NSABB Subcommittee: FASEB's policy development process occurs through its Science Policy Committee, which functions through subcommittees or working groups of experts. In response to the U.S. National Science Advisory Board on Biosecurity's proposed oversight framework, FASEB formed a subcommittee to address dual use issues generally and the NSABB proposal specifically. The chair of that subcommittee, Dr. Avrum Gotlieb, participated in the November dual use meeting organized by the Polish Academy of Sciences (see above). The subcommittee and staff continue to monitor and respond to dual use issues as they arise and develop policy statements as appropriate. FASEB has worked to raise awareness of dual use research issues through periodic publications in society newsletters, as well as our own electronic newsletter.

Related Activities: We have surveyed the FASEB leadership and membership about dual use research issues and have found very low levels of awareness. Respondents suggested that scientific meetings would serve as a valuable outreach tool, although this conflicts with our experience. Typically, the attendance at policy sessions during society meetings is fairly low. Moreover, FASEB member societies have their own priorities for the limited policy sessions at scientific meetings and dual use research was identified as a low policy priority. FASEB has supplied a number of informative articles on dual use research and the activities of NSABB for society newsletters, as well as our own electronic newsletter, the *FASEB Washington Update.* In addition, FASEB has been actively engaged individually and with coalition partners in policy development on a number of related issues, including deemed exports, visa issues, and Select Agent regulations.

THE INTERACADEMY PANEL ON INTERNATIONAL ISSUES (IAP)

The IAP, founded in 1993, is a global network of 100 science academies in partnership designed "to help its members develop the tools they need to participate effectively in science policy discussions and decision making." The current co-chairs are Chen Zhu (Minister of Health, China) and Howard Alper (RSC: The Academies of Arts, Humanities and Sciences of Canada). More information can be found on the IAP Web site at *http://www.interacademies.net/.* The IAP Executive Council established a Biosecurity Working Group (BWG) in 2004 to coordinate its activities in this area; its members are the academies of China, Cuba, the Netherlands (chair), Nigeria, the United Kingdom, and the United States. The BWG has undertaken a number of activities related to dual use issues.

In March 2005, the IAP, the International Council for Science (ICSU), the InterAcademy Medical Panel (IAMP) and The National Academies of the United States hosted the International Forum on Biosecurity in Como, Italy. The Forum was designed to serve as a convening and coordinating mechanism to share information about activities under way or being planned and to broaden the debate and advance the awareness in the life sciences and biomedical research communities—and in the international scientific community more generally—about the challenges posed by the "dual use" dilemma.

In December 2005, the IAP released a Statement on Biosecurity, which has been endorsed by over 70 national science academies. The statement provides principles for academies and other scientific bodies preparing codes of conduct that address five fundamental issues facing scientists working in the biosciences—awareness; safety and security; education and information; accountability; and oversight.

In September 2006, IAP, ICSU, and the Royal Society hosted the workshop Scientific and Technological Developments Relevant to the Biological and Toxin Weapons Convention. The workshop brought together 84 scientific and policy experts from 23 countries to consider recent developments in the biosciences and their potential implications. A statement and report were produced from the meeting that aimed to inform delegates at the Sixth Review Conference of the BWC.

THE INTERACADEMY MEDICAL PANEL (IAMP)

The InterAcademy Medical Panel, a global network of academies of science and medicine, is committed to improving health world-wide. Currently the IAMP has 64 members; more information can be found on its Web site (*http://www.iamp-online.org/*). The current co-chairs are Guy de Thé, Académie de Médicine, France, and Anthony MBewu, Academy of Sciences of South Africa. Its activities focus on institutional collaboration to strengthen the role of all academies to alleviate the health burdens of the world's poorest people; build scientific capacity for health; and provide independent scientific advice on promoting health science and health care policy to national governments and global organizations.

THE INTERNATIONAL UNION OF BIOCHEMISTRY AND MOLECULAR BIOLOGY (IUBMB)

The International Union of Biochemistry and Molecular Biology—founded in 1955—unites biochemists and molecular biologists in 66 countries that belong to the Union as Adhering or Associate Adhering Bodies, representing biochemical societies, national research councils, or academies of sciences. The IUBMB is devoted to promoting research and education in biochemistry and molecular biology throughout the world and gives particular attention to areas where the subject is still in its early development. It achieves this in several ways.

Every three years the IUBMB sponsors an International Congress of Biochemistry and Molecular Biology. Cosponsorship of these Congresses by Regional Organizations of Biochemistry and Molecular Biology is an increasing trend. These Congresses are major international meetings where current research in all fields of biochemistry and molecular biology is considered. Thousands of individual research projects are presented in poster sessions and leading investigators from many nations survey their fields and describe their own research in symposia and plenary lectures. Since 1992 IUBMB has also sponsored IUBMB Conferences and Special Meetings, held in the years between the International Congresses.

The IUBMB provides financial support for international symposia on biochemical and molecular biological research topics of current interest. It organizes or sponsors workshops, symposia and training sessions on biochemical and molecular biological education and provides free textbooks and journals to training institutions in developing nations. The IUBMB also funds short-term fellowships for younger biochemists and molecular biologists to travel to other institutions to perform research not possible in their own laboratories, and provides Travel Fellowships for young scientists to attend its Congresses. Sponsorship of meetings and fellowships is restricted to regions that belong to the IUBMB.

As well as reaching biochemists through its own meetings, the IUBMB works closely with the four regional organizations that unite the biochemical societies of Asia and Oceania (Federation of Asian and Oceanian Biochemists and Molecular Biologists), Europe (Federation of European Biochemical Societies) the Americas (Pan-American Association for Biochemistry and Molecular Biology) and Africa (Federation of African Societies of Biochemistry and Molecular Biology). Indeed all four are linked formally with the IUBMB as Associated Regional Organizations and three of them receive substantial financial support from the IUBMB. The International Federation of Clinical Chemistry and Laboratory Medicine, the International Society for Neurochemistry, the International Organization for Free Radical Research, and the International Society of Vitamins and Related Biofactors are also Associated Organizations of IUBMB.

Reaching individual biochemists is also the purpose of another very important function of the IUBMB, that of publishing news, reviews, information, original research, and nomenclature. *Trends in Biochemical Sciences* (*TiBS*) is seen monthly by over 100,000 readers, keeping them informed of research progress across the broad field of biochemistry and molecular biology, as well as of news of meetings, people and biochemical events. *Biotechnology and Applied Biochemistry* publishes original research findings and reviews in the expanding domain of the practical applications of the subject. *IUBMB Life* expedites the publication of short communications, identified by their novelty and the need for urgent dissemination. *Biochemistry and Molecular Biology Education* is dedicated to publishing articles, reviews and editorials to assist the teaching of biochemistry and molecular biology to science and medical students throughout the world. *BioEssays*, cosponsored by the IUBMB and seven other ICSU biological Unions, is the monthly current-awareness journal that displays progress across the fields of molecular, cellular and developmental biology. *BioFactors* publishes reviews and original communications on growth factors and regulatory substances. *Molecular Aspects of Medicine* publishes reviews that aim to link clinicians and biomedical scientists.

The IUBMB is one of 29 Scientific Unions affiliated with the International Council of Science, an umbrella organization for scientists worldwide. ICSU was created in 1931 to encourage international scientific activity, to affirm the rights of scientists without regard to race, religion, political philosophy, ethnic origin, sex or language to join in international scientific affairs for the benefit of mankind. The IUBMB has been a member of ICSU since 1955 (until 1991 as IUB). The IUBMB representative serves as a member of the General Assembly of ICSU and ex-officio takes part in the work of the ICSU working group of the Biological Sciences. Through ICSU the IUBMB has been able to generate broad and often highly productive contacts with other international bodies, including some joint programs.

Further information is available online at *www.iubmb.org*.

THE INTERNATIONAL UNION OF BIOLOGICAL SCIENCES (IUBS)

The International Union of Biological Sciences is a non-governmental, non-profit, scientific network founded in 1919. The membership of IUBS presently consists of 44 Ordinary Members, adhering through Academies of Sciences, National Research Councils, national science associations or similar organizations, as well as 80 Scientific Members, all of which are international scientific associations, societies or commissions focusing on a wide array of biological disciplines. IUBS was one of the founding unions of the International Council for Science, and IUBS continues to contribute to the work of ICSU's scientific committees and programs. The objectives of the IUBS are: to promote the study of biological sciences; to initiate, facilitate and coordinate research and other scientific activities necessitating international, interdisciplinary cooperation; to ensure the discussion and dissemination of the results of cooperative research, particularly in connection with IUBS scientific programs; and to support the organization of international conferences and assist in the publication of their reports. IUBS organizes triennial General Assemblies, which are flanked by a scientific symposium organized in cooperation with the National IUBS Committee of the host country. It also conducts scientific programs, which currently include Biological Diversity, Integrative Biology, Biological Education, Bioethics, Integrative Climate Change Biology, Bio-Energy, Biology & Traditional Knowledge and the 2009 Darwin Celebration Year. IUBS publications include the quarterly periodical *Biology International*, the IUBS Monograph Series, the Methodology Manual Series and the *Proceedings of the General Assemblies*.

Further information is available online at *www.iubs.org*.

THE INTERNATIONAL UNION OF MICROBIOLOGICAL SOCIETIES (IUMS)

The International Union of Microbiological Societies is one of the 29 Scientific Unions of ICSU. It was founded in 1927 as the International Society of Microbiology, and became the International Association of Microbiological Societies affiliated with the International Union of Biological Sciences (IUBS) as a division in 1967. It acquired independence in 1980 and became a Union Member of ICSU in 1982. IUMS currently has 113 member societies and 14 associate members representing well over 100 countries. Members are National Societies and Associations for Microbiologists and associate members are other institutions with an interest in microbiological and connected sciences.

The objectives of the Union are to promote the study of microbiological sciences internationally: initiate, facilitate and coordinate research and other scientific activities that involve international cooperation; ensure the discussion and dissemination of the results of international conferences, symposia and meetings and assist in the publication of their reports; represent microbiological sciences in ICSU; and maintain contact with other international organizations.

The major goal of IUMS is to promote research and the open exchange of scientific information for advancement of the health and welfare of humankind and the environment and strongly discourages any uses of knowledge and resources to the contrary. In particular, the IUMS strives to promote ethical conduct of research and training in the areas of biosecurity and biosafety so as to prevent use of microorganisms as biological weapons and therefore to protect the public's health and to promote world peace. IUMS seeks that all its member societies adopt or develop a Code of Ethics to prevent misuse of scientific knowledge and resources. The IUMS Code of Ethics Against Misuse of Scientific Knowledge, Research and Resources is available from the IUMS Web site at *http://www.iums.org/*

about/about_us-Codeethics.html. The Code has been approved by the Executive Board and the approval of the member societies has been requested.

Further information is available online at: *www.iums.org*.

THE INTERNATIONAL UNION OF PURE AND APPLIED CHEMISTRY (IUPAC)

The International Union of Pure and Applied Chemistry serves to advance the worldwide aspects of the chemical sciences and to contribute to the application of chemistry in the service of humankind. As a scientific, international, nongovernmental and objective body, IUPAC can address many global issues involving the chemical sciences.

IUPAC was formed in 1919 by chemists from industry and academia. Over nearly nine decades, the Union has succeeded in fostering worldwide communications in the chemical sciences and in uniting academic, industrial and public sector chemistry in a common language. IUPAC has long been recognized as the world authority on chemical nomenclature, terminology, standardized methods for measurement, atomic weights and many other critically evaluated data. The IUPAC continues to sponsor major international meetings that range from specialized scientific symposia to CHEMRAWN (CHEMical Research Applied to World Needs) meetings with societal impact. During the Cold War, IUPAC became an important instrument for maintaining technical dialogue among scientists throughout the world.

IUPAC is an association of bodies, National Adhering Organizations, which represent the chemists of different member countries. There are 45 National Adhering Organizations, and 20 other countries are also linked to IUPAC in the status of Associate National Adhering Organizations. Almost 1,000 chemists throughout the world are engaged on a voluntary basis in the scientific work of IUPAC, primarily through projects, which are components of eight divisions and several other committees.

Further information is available online at: *www.iupac.org*.

THE INTERNATIONAL UNION FOR PURE AND APPLIED BIOPHYSICS (IUPAB)

The International Union for Pure and Applied Biophysics is a member of the ICSU family. Affiliated to it are the national adhering bodies of 50 countries. Its main objectives are to support research and teaching in biophysics, promote communication between the various branches of biophysics and allied subjects, and to encourage cooperation between the societies that are interested in the advancement of biophysics in all of its aspects.

In order to achieve these objectives, the Union organizes triennial International Congresses and General Assemblies, which will next be held in China in 2011. IUPAB has four Task Forces concerned with major areas of biophysics: Bioinformatics, Capacity Building and Education in Biophysics, Nuclear Magnetic Resonance in Biological Sciences, and Biomedical Spectroscopy. The Task Forces also arrange specialist meetings either associated with the Congresses or, more commonly, in the intervals between Congresses. The Union also supports conferences, schools and workshops, with priority given to events that will promote biophysics in the developing countries and that will facilitate the participation of young scientists in the conferences that it supports.

Further information is available online at *www.iupab.org*.

THE NATIONAL ACADEMIES

The National Academies of the United States comprises the National Academy of Sciences (NAS), the National Academy of Engineering (NAE), the Institute of Medicine (IOM) and the National Research Council (NRC). Some of the most relevant international Biosecurity work includes:

- The International Biosecurity Project works to promote implementation of the international recommendations of the 2004 NRC report *Biotechnology Research in an Age of Terrorism.* A collaboration among several units at the National Academies, the project's overarching goal is to develop and promote more effective international strategies to reduce the risk that advances in life sciences research could be misused. A key element involves working with international partners other academies and international scientific organizations, as well as a wide range of intergovernmental and nongovernmental organizations.
- The Biological Threats Panel brings together National Academy of Sciences Committee on International Security and Arms Control (CISAC) and non-CISAC experts to address the scientific and technical dimensions of biological weapons, bioterrorism, issues related to successful implementation of the Biological and Toxin Weapons Convention, biosecurity, and other contemporary challenges related to rapid growth in biotechnology. The Panel coordinates across the Academies with ongoing efforts and develops its activities in partnership programs inside and outside the Academies. The Biological Threats Panel continues work started in 1986 by CISAC's Biological Weapons Working Group (BWWG), whose initial focus was on continuing concerns about Soviet compliance with the Biological Weapons Convention. In the mid-1990s the BWWG played a leading role in fostering U.S. government support for cooperative research programs between American scientists and scientists from former Soviet

biological weapons research institutes. Recently CISAC's Biological Threats Panel has established counterpart groups through the Russian Academy of Sciences CISAC and with the Chinese Biological Scientist's Group of the Chinese People's Association of Peace and Disarmament.

- The Board on International Scientific Organizations (BISO) examines issues related to the conduct of science, evaluates opportunities for international collaboration in scientific research, and strengthens U.S. participation in international scientific, engineering, and medical organizations. The Board also oversees a network of more than 20 U.S. national committees corresponding to ICSU scientific member bodies, seeks committee input on issues confronting ICSU and its bodies, and informs them of the input NAS is considering in its role as a national member of ICSU. Scientific unions in the biological and chemical sciences with which BISO is involved include IUBS, IUBMB, IUMS, IUPAB, IUPAC, and others.

Further information about The National Academies can be found at *http://nationalacademies.org/*. Information about its work in biosecurity can be found at *http://www7.nationalacademies.org/biosecurity/*.